THE SKEPTIC'S GUIDE TO SPORTS SCIENCE

The global health and fitness industry is worth an estimated $4 trillion. We spend $90 billion each year on health club memberships and $100 billion each year on dietary supplements. In such an industrial climate, lax regulations on the products we are sold (supplements, fad-diets, training programs, gadgets, and garments) result in marketing campaigns underpinned by strong claims and weak evidence. Moreover, our critical faculties are ill-suited to a culture characterized by fake news, social media, misinformation, and bad science. We have become walking, talking prey to 21st-century Snake Oil salesmen.

In *The Skeptic's Guide to Sports Science*, Nicholas B. Tiller confronts the claims behind the products and the evidence behind the claims. The author discusses what might be wrong with the sales pitch, the glossy magazine advert, and the celebrity endorsements that our heuristically wired brains find so innately attractive. Tiller also explores the appeal of the *one quick fix*, the fallacious arguments that are a mainstay of product advertising, and the critical steps we must take in retraining our minds to navigate the pitfalls of the modern consumerist culture.

This informative and accessible volume pulls no punches in scrutinizing the plausibility of, and evidence for, the most popular sports products and practices on the market. Readers are encouraged to confront their conceptualizations of the industry and, by the book's end, they will have acquired the skills necessary to independently judge the effectiveness of sports-related products. This treatise on the commercialization of science in sport and exercise is a *must-read* for exercisers, athletes, students, and practitioners who hope to retain their intellectual integrity in a lucrative health and fitness industry that is spiraling out of control.

Nicholas B. Tiller is a research fellow in Exercise Physiology, at Harbor-UCLA Medical Center. He was born in London, England. Tiller holds master's and doctoral degrees in Human Applied Physiology, and is accredited with the British Association of Sport and Exercise Sciences (BASES). He writes about science, health, exercise, and critical-thinking, and is an avid ultra-marathon runner.

THE SKEPTIC'S GUIDE TO SPORTS SCIENCE

Confronting Myths of the Health and Fitness Industry

Nicholas B. Tiller

NEW YORK AND LONDON

First published 2020
by Routledge
52 Vanderbilt Avenue, New York, NY 10017

and by Routledge
2 Park Square, Milton Park, Abingdon, Oxon, OX14 4RN

Routledge is an imprint of the Taylor & Francis Group, an informa business

Library of Congress Cataloging-in-Publication Data
Names: Tiller, Nicholas B., author.
Title: The skeptic's guide to sports science: confronting myths of the health and fitness industry / Nicholas B. Tiller.
Description: New York: Routledge, 2020. |
Includes bibliographical references and index. |
Identifiers: LCCN 2019052968 (print) | LCCN 2019052969 (ebook) |
ISBN 9781138333123 (hardback) | ISBN 9781138333130 (paperback) |
ISBN 9780429446160 (ebook)
Subjects: LCSH: Sports sciences—Social aspects. | Physical fitness—Social aspects. |
Sporting goods industry—Marketing. | Sport clothes industry—Marketing. |
Dietary supplements industry—Marketing. | Physical fitness centers—Marketing. |
Personal trainers—Marketing. | Deceptive advertising.
Classification: LCC GV558 .T55 2020 (print) |
LCC GV558 (ebook) | DDC 613.7/1—dc23
LC record available at https://lccn.loc.gov/2019052968
LC ebook record available at https://lccn.loc.gov/2019052969

ISBN: 978-1-138-33312-3 (hbk)
ISBN: 978-1-138-33313-0 (pbk)
ISBN: 978-0-429-44616-0 (ebk)

Typeset in Bembo
by codeMantra

'For the skeptical movement'.

CONTENTS

FIGURES

PREFACE

In the relative infancy of our species, evolutionary pressure imprinted on our brains an ethos of economy. This hardwired behavior served us well in navigating hyper-social groups, mitigating unnecessary risk, and avoiding starvation by preserving energy in times of calorific scarcity; we endured by taking shortcuts and following a path of least resistance. But we're conditioned with an economy ill-suited to the modern world. In our contemporary society, characterized by a surplus of calories (energy) and all the contrivances life can afford, our misplaced behaviors have birthed an obesity epidemic we cannot stem, and a commercialist society which feeds on our desire for quick fixes and simple resolutions to complex problems. Although considered the rational animal, the modern environment has changed faster than our logic and reason have been able to adapt. Our ancestors never had to contend with fake news, misinformation, bad science, or social media. A poor understanding of science and a dearth of critical-thinking are exploited in the sale of products we often don't need. We are walking, talking prey to the 21st-Century Snake Oil salesman.

These New-Age arbiters of speed, strength, power, weight-loss, health, vitality, and wellbeing offer us shortcuts by way of shoes, socks, sports drinks, tights, training programs, wristbands, garments, powders, pills, and potions. Scarcely are the claims for efficacy – the adverts in glossy magazines, websites, and television commercials – made on solid ground, on a foundation of legitimate authority, supported by scientific evidence. They often fail the test of close scrutiny. In the ongoing war between rationality and irrationality, tackling the *pseudo*science of sport is a principal battle. It's the focus of this book because it unifies two of my passions: sports science and scientific skepticism. I studied the former as an undergraduate at the University of Hertfordshire, UK, where I developed a fascination with exercise physiology. I vividly recall a lecture on aviation physiology in which we learned about the wonderfully complex physical reactions to environmental stimuli. It was then I realized my desire for a career in science. As a body of knowledge, science is a catalogue of human endeavor. It affords us the capacity to understand, at least in part, the world around us, removing much of the guess-work from our existence. The scientific process, gradually stripping away life's mysteries, bringing light to the dark, has explained everything from the origin of the human species to the behavior of fundamental particles, and much of the puzzle in between. And while there are many questions for which science yet has no answer, we've made incredible progress in our short time as the rational animal.

When compared to other disciplines, the exercise sciences (to include physiology, nutrition, sports psychology, physiotherapy, strength and conditioning, biomechanics) are young endeavors, and some of our guiding principles are less well-established, in both academic and applied circles. As such, it's become a breeding ground for myth and misinformation, its concepts exploited for monetary gain. The sale of sports products – the supplements, garments, trainers, fad diets, and exercise programs – is poorly regulated by the federal agencies that are appointed to protect consumer interests. Many products have meager supporting evidence-for-efficacy, sold on false claims and bad science. The result is a consumer market dominated by quacks and charlatans that thrive in a post-truth era where facts are subordinate to emotions, morality subordinate to money, and science subordinate to marketing. The difficult challenge of successfully navigating such a climate is compounded by an educational system which often lacks formal instruction on critical-thinking. When I qualified as a practitioner (a physiologist), I was ill-equipped to tackle the commercialist world around me, let alone advise others on the path. It's only in retrospect that I reflect on my naiveté; I was ignorant of the reach of my own ignorance.

That changed when my journey into skepticism began in 2006. I'd studied the tenets of critical-thinking and owned several books on logic, but my fortuitous discovery of *The Skeptic's Guide to the Universe* – a science and critical-thinking podcast – proved to be my gateway to an entire community of scientists and critical-thinkers devoted to the cause. The further down the rabbit hole I ventured, the more wonderous it became and the more knowledge I acquired. The term scientific skepticism – coined by renowned astrophysicist, astronomer, and writer Carl Sagan – isn't a doctrine or set of rules, and isn't concerned with instruction on what to think, but rather how to think. Above all, it's the practice of questioning whether claims are supported by plausibility and empiricism, pursuing only the extension of certified knowledge. Skeptics place evidence above all conclusions that might result, and I consider this to be an essential practice for anyone in search of fact over fiction. The more I studied the ethos of skepticism, the more it became apparent that its basic principles of honesty, truth, and humility were in stark contrast to those of the health and fitness industry.

In this book, we take an objective look at the myriad health and performance claims that stem from the fitness industry and, in doing so, I'll discuss the concepts of scientific skepticism and their relevance to the world of sports science. As we progress, you'll acquire the essential skills you need to judge for yourself the efficacy of sports products and practices, so that never again will you be deceived by flawed arguments and deceptive marketing. The result will be the ability to choose strategies from which you're likely to benefit, whatever your goals, while avoiding the frustration that comes from squandered time and money. Throughout the main text, we'll look at the scientific plausibility of, and evidence for, numerous health and sports products, debunking many of them in the process. We'll explore the appeal of the *one quick fix*, the fallacious arguments that are the lynchpin of product advertising, and the critical steps we must take in retraining our minds to navigate the pitfalls of our modern consumerist culture. There will be some basic physiology, but don't let that put you off. And although this is a book about sports science, critical-thinking, and the commercialization of the health and fitness industry, at its core it's a book about how to uncover the truth.

It's my hope that by the end of this modest work, you'll have a greater appreciation of the various manifestations of *pseudo*science in sport, but you'll also have developed the critical faculties you need to arm yourself in the ongoing fight for the truth about health and performance. I also aspire to spark or invigorate a sincere appreciation of the true knowledge and wonder that science can afford. As a system of thought, scientific skepticism has fundamentally changed the way I view the world, and in writing this book I hope to give something back to a community from which I've already taken so much.

Nicholas B. Tiller
September 2019

ACKNOWLEDGMENTS

I'll spare the reader from a directory of acknowledgments because, in all candor, I didn't receive a great deal of assistance in producing the manuscript. Nevertheless, a select few individuals played a crucial role in its progression. My good friend Andrew gave me his unwavering support from the book's conception to publication; several of the arguments herein came from his unrivaled keenness and alacrity of mind. I owe a debt of gratitude to Prof. Simon Shibli, a respected scholar and thoroughly decent individual, one of the few senior academics using his influence to hearten and support those striving to ascend the ranks. I'd like to thank David Varley, Megan Smith, and Simon Whitmore of Routledge for having faith in the book's premise, and in me to deliver on it. Finally, while I can't thank him, I'd like to acknowledge my late father who never for a second doubted that the manuscript would see its conclusion. His meditations on when (not if) the book was going to be published were a constant source of inspiration.

1

SNAKE OIL FOR THE 21ST-CENTURY

"Some myths deserve to be broken, out of respect for the human intellect".

– Neil deGrasse Tyson

com·mer.cial·ism.
An attitude that emphasises tangible profit or success.

1.1. Beginnings

It was the summer of 1864, on the Western coast of the United States. Irish and Chinese immigrants, along with a handful of Union and Confederate military veterans, had long since risen from a brief and broken sleep to resume their backbreaking labors. In the relentless blaze of the scorching sun, workers connected the steel rails that would eventually form the 3,000-km stretch of track comprising the First Transcontinental Railroad, later known as The Overland Route. This continuous railway, the first of its kind, would eventually travel the entire length of the Western United States to connect San Francisco with Iowa. Forging this iconic railway was dangerous and exhausting work. Indeed, the track would eventually comprise around 750,000 rails connected end-to-end, each one fixed in position with ten metal spikes, each spike needing several blows from a 12–16 lb spike maul to penetrate the dry, barren earth. The greatest threat to the lives that made possible this arduous feat was not the 30 million hammer swings, nor the punishing hours of work, dehydration, or poor sanitation; the conspicuous danger of demise came from the Sierra Nevada Mountains and the millions-of-tons of ancient rock precluding connection of the Donner Pass. After railway progress had ground to a halt, the decision was made to explode and remove 500 m of dense mountainous rock using Nitroglycerin. Due to its severe instability, *Nitro* couldn't be safely (or legally) transported to the labor camps, and so a Scottish chemist named

James Howden offered to mix it on-site. An estimated 1,500 Chinese workers died in explosions or subsequent rock slides.

The long days of grueling work caused great physical stress which punished the bodies of the railway workers. As an elixir for the pain, the Chinese laborers – who were mostly responsible for building bridges and tunnels – would massage oil derived from the Chinese water snake into their muscles and joints. They believed that this ancient remedy – passed down through the generations – contained magical properties that would heal their dilapidated limbs. They shared the tonic with their American counterparts who, seeing great lucrative potential in the tonic, began selling it throughout North America, eventually touring the country to promote various Snake Oil preparations. Before long, sales had proliferated into a booming business, helped in part by traveling salesman who staged theatrical performances for gathering onlookers. During such performances, an accomplice of the salesman would be planted in the gathering crowd in the guise of a customer. The accomplice, feigning a physical ailment, would use the Snake Oil preparation before dramatically proclaiming its miraculous benefits. Pervasive, too, in contemporary culture, such anecdotal testimonies held powerful dominion over the consumers deceived into the purchase. Be cognizant that, despite emerging at a fairly advanced stage of the scientific revolution, the Old West lacked the formal means to refute the grand claims that Snake Oil was a powerful pain reliever. Subsequently, by the end of the 19th-Century, Snake Oil was feverishly popular due primarily to the relative scientific illiteracy of the average customer.

In the 1900s, there were major developments in analytical chemistry – the study of the chemical components of materials – and in 1916, the Western Snake Oil preparation was scrutinized by the United States Government. The Secretary of Agriculture penned a damming report; according to the US District Court, Snake Oil contained mineral oil, beef fat, black pepper (to warm the skin), turpentine or camphor (to produce a medicinal smell), but crucially, no actual Snake Oil. Clark Stanley, the product manufacturer of the time, was prosecuted by the US government and fined $20 for misrepresenting and selling a product with no active ingredient, while simultaneously claiming that his preparation was *the strongest and best liniment known for pain and lameness.* Alongside traditional Snake Oil, other ineffective nostrums like Powdered Unicorn Horn were sold as cure-alls well into the 1900s, predominantly to customers who'd been duped into perceiving powerful healing properties (Figure 1.1).

Although Stanley's concoction was a proven sham, it seems unlikely that an earlier formula was any more effective as a method of pain relief. To this day, there's no convincing evidence that Snake Oil – a source of eicosapentaenoic acid – is effective for pain relief when either consumed or applied directly to the skin. Snake Oil is just one example of ancient Chinese medicine which, over the centuries, has birthed hundreds of products (like turtle shell, shark fin, beetles, tiger penis, and dried human placenta) all claiming miraculous

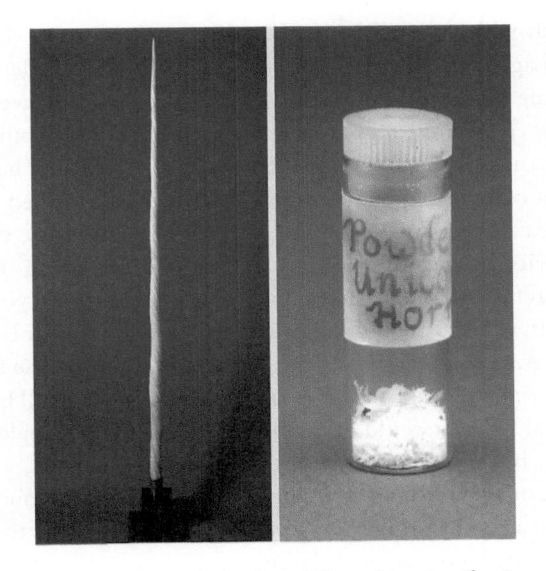

FIGURE 1.1 Alongside traditional Snake Oil, other ineffective nostrums like Powdered Unicorn Horn were sold as cure-alls well into the 1900s. Unsurprisingly, we're still waiting for the first controlled studies on the effects of this elusive panacea.

Source: Image courtesy of the U.S. National Library of Medicine.

health benefits. Moreover, a principal cause of poaching (of animals like the African Rhino which has been hunted to near-extinction for its ivory tusk) is the superstitious beliefs propagated by ancient Chinese medicine and other *pseudo*sciences. The term *Snake Oil* is now synonymous with false claims and theatrical pageantry, but not medicine. It's a product with questionable or unverifiable qualities, offering unrealistic outcomes, without any supporting evidence-for-efficacy.

Snake Oil salesmen still exist today; they sell speed, strength, power, weight-loss, health, vitality, and wellbeing, in shoes, socks, sports drinks, tights, training programs, wristbands, garments, powders, pills, and potions. The shortcuts they offer – quick fixes to health and sporting prowess – are packaged and sold with a theatrical ad-campaign and shrewd marketing rhetoric. A Snake Oil salesman is someone who sells goods on false or exaggerated claims, and the sports industry is rife with them.

1.2. Why Do We Take Shortcuts?

The fact that you're reading this book suggests that at some point, like me, you've *bought* a *product* in the hope that it'll benefit your health or sports performance; however, it's important that I first clarify these terms given the frequency with which they'll be used in the remaining text. The verb *bought* in its formal usage is the past and past participle of *buy*, as in *obtained in exchange for payment* using money or goods. In its informal usage, however, *bought* could just mean to *accept the truth of.* For example, before you buy some compression tights on the basis that they might facilitate recovery, you first have to buy into the notion that compression garments facilitate recovery. When a fitness coach claims that she can help you lose 15 lbs for your beach holiday, you can choose to buy her advice in exchange for money but, prior to doing so, you'll likely buy into the premise that she knows how to help you achieve your goals in the first instance. In either case, buying can be expensive. Be cognizant, therefore, of the various ways this term is invoked. A *product* is *an article or substance that is manufactured or refined for sale.* In this book, the word *product* will be employed in its more general sense; it could be a tangible 'thing' like a supplement or some running shoes, but it could also be used to denote a fad diet, a training program, a brand, or something altogether more abstract like a concept or theory. In other words, a product is literally anything that can be sold to the consumer on the basis of a claim or series of claims.

Now, think back to the last time you bought a product. It might have been a nutritional supplement like a multivitamin, or perhaps you own an activity or heart rate monitor, or some high-tech running shoes, or you invested in ten sessions of personal training at your local gym, or you believed the personal trainer when he assured you that his personalized nutrition strategy would grant you 10 lbs of muscle in three weeks. It'd be a dramatic oversimplification of behavioral psychology to suppose that these products hold an appeal simply because we care about our health and fitness. Before we can learn how to make good decisions in this commercialist climate, we first need to understand the motivations and biases that inform them. We each have our own personal reasons for making such choices, but there are several common, interrelated reasons why we seek shortcuts to health and performance.

A heuristic (noun) is a cognitive process, a type of mental shortcut, which for the sake of economy obligates you to ignore certain bits of information. Harnessing heuristics in decision-making ensures a practical approach – sufficient for obtaining an outcome – but one that's not guaranteed to be optimal or even correct. Heuristic decisions save effort, but were traditionally thought to result in greater errors than the so-called *rational* decisions that are predicated on logic. These mental shortcuts are characterized by the engrained human yearning for brevity, itself attributable to evolutionary pressure and sociocultural factors. In essence, heuristics are a collection of naturally developed mental strategies that shorten decision-making time to help us seek out a path of least resistance; it's largely responsible for the popularity of Snake Oil and its descendants. From an evolutionary standpoint, we are hard-wired for heuristic thinking because it once served a critical survival advantage. Our extinct human ancestors (and most of the existing ape species alive today) depended heavily on their ability to forage and hunt for food, and heuristic thinking helped them to conserve energy in times of calorific scarcity. Those organisms exhibiting the greatest economy inevitably found more food and were more likely to survive and propagate offspring; subsequently, their heuristically inclined genes would be propagated. In a fascinating study published in the journal *Animal Behaviour* in 2012[1], chimpanzees were observed implementing heuristic problem-solving in their search for food. The researchers packaged almonds in a range of quantities, sometimes shelled and sometimes unshelled, and buried them at various locations in a large wooded area. When encouraged to forage, the chimpanzees sought out almonds first by quantity, then by shell presence or absence, and finally by distance from home. The recovery sequence was closely related to the energy/handling time profitability of the food source, and reflected the most economical foraging strategy. It's an extraordinary insight into the primal decision-making abilities of our closest cousins, faculties that are very much a part of human DNA.

This economy can still serve a purpose in aspects of modern culture. In business organizations, healthcare, and legal institutions, ignoring certain data points in the decision-making process can lead to more accurate judgments when compared to weighing-up *all* of the information available[2]. In medicine, for example, heuristics can be used to make fast, accurate, and cost-effective decisions[3]. The pitfall, however, is that knowing what information is relevant and what can be ignored requires knowledge, training, and experience; identifying the key predictors of an outcome requires a great deal more than intuition. In some instances, therefore, heuristic thinking can lead to poor decisions and tremendous bias. Consider our contemporary food culture. We no longer exert a great deal of energy in securing our meals. Instead of expending calories by foraging and hunting, we buy our groceries pre-packaged at the local store. If this is too time-consuming for our busy schedule, we can have groceries delivered straight to our front door and, if we're not in the mood to prepare them, we can order a takeaway and wait 30 minutes for plastic containers of piping-hot, calorie-dense food to appear on our laps. In securing the meal, we have only expended the energy necessary to open the refrigerator, dial the phone, open a smartphone application, or answer the door. With so much energy available at all hours of the day or night in exchange for meager effort, our society is one that encourages calorific excess, leading to a population in which one in four people (one in three in the US) is clinically obese. This is one example of economical decision-making ill-suited to the current climate.

Some thought, therefore, must be given to the appropriateness with which we rely on our inherent human faculties – or intuition – when making decisions. This is a concept that's explored later in more detail. And while for some decisions our intuition can serve us very well, there are others for which intuition alone is insufficient for achieving an appropriate outcome. We must strive to engage in more cognitive, thoughtful decisions. In *Mysticism and Logic*, Bertrand Russell suggests:

> Where instinct and reason do sometimes conflict is in regard to single beliefs, held instinctively, and held with such determination that no degree of inconsistency with other beliefs leads to their abandonment. Instinct, like all human faculties, is liable to error. Those in whom reason is weak are often unwilling to admit this as regards to themselves, though all admit it in regard to others.

When we follow the path of heuristics in the world of health and fitness, we're led inevitably to the ergogenic aid; this is anything external to the body that's deemed to directly or indirectly enhance performance or training adaptation, and its existence is fundamental to the industry. Ergogenics can be loosely categorized into four branches: (1) mechanical aids: these are objects that have an internal mechanism, or that in some way inform the manner in which we train. Some examples are heart rate monitors, power meters for your bike, running shoes, and joint supports; (2) psychological aids: these influence the way we think, the way we feel, and our perceptions and experiences of exercise. Some examples are imagery, a cheering crowd, the placebo effect, and music; (3) physiological aids: ergogenics that help us to directly manipulate physiological function. These could include physiological profiling tools like a fitness test, altitude training, blood doping, and heat acclimation which might enhance performance when a competition is contested in a hot and/or humid environment; (4) nutritional aids: this is dietary manipulation to influence physiological function. Some examples are carbohydrate loading before a marathon, taking supplements like protein, or drugs like caffeine. There is substantial gray area in distinguishing between physiological and nutritional aids which warrants a brief exploration. Nutritional supplements (most of which are taken orally) are generally considered to be the domain of the sports nutritionist, but this engenders a problem when we note that not all performance-enhancing supplements are taken orally. Testosterone – the illegal anabolic steroid – can be administered orally (in which case it enters the blood via the GI tract), but can also enter the blood via intramuscular injection, or transdermal patch worn on the skin; as a result, it cannot be labeled exclusively as a nutritional aid. It doesn't seem intuitive to consider oral supplements as nutritional ergogenics and non-oral supplements as physiological ergogenics. And what about medical drugs like orally ingested pain-killers which are performance-enabling; it would hardly be appropriate for these to be considered the domain of a nutritionist. And oxygen taken in via the mouth; should that also be considered among the repertoire of a nutritionist? Accordingly, oral ingestion is not a sufficient criterion for the classification of a nutritional ergogenic aid. In a recently updated review[4], the *International Society of Sports Nutrition* offered this clarification specific to dietary supplements:

> The law defines a "dietary supplement" as a product that is intended to supplement the diet and contains a "dietary ingredient". By definition, "dietary ingredients" in these products may include vitamins, minerals, herbs or other botanicals, amino acids, and substances such as enzymes, organ tissues, and glandular extracts. Further, dietary ingredients may also include extracts, metabolites, or concentrates of those substances. Dietary supplements may be found in many forms such as tablets, capsules, softgels, gelcaps, liquids, or powders, but may only be intended for oral ingestion. Dietary supplements cannot be marketed or promoted for sublingual, intranasal, transdermal, injected, or in any other route of administration except oral ingestion.

Congruent with their suggestion, and to accommodate the discussions in this book, I will define a nutritional ergogenic as anything which offers caloric and/or nutritional value, i.e., proteins, carbohydrates, fats, and the micronutrients. Everything else will be considered a physiological ergogenic. Playing devil's advocate, one could suppose that all interventions are designed for the purpose of manipulating physiological function to some end. In any case, humans have been using these shortcuts for many millennia, and our interest in these ergogenic aids is attributable, at least in part, to our engrained propensity for heuristic thinking.

Another explanation for the popularity of the ergogenic aid is that contemporary society has accustomed us to treating *symptoms* rather than *causes*. To lose weight (or more specifically, fat), we're compelled first to fat-burning supplements or crash weight-loss diets rather than taking a holistic look at our eating habits, our relationship with food, or increasing our physical activity; we treat the symptoms of being overweight instead of the causes. Many deal with injury the same way. As a student, I had a weekend job in a sports store selling running shoes and performing rudimentary biomechanical analyses on running gait. Runners would occasionally present with muscle soreness or niggling injuries caused by arduous training and insufficient rest, but instead of assessing their underlying strength imbalances, or loading issues, or getting more sleep, or improving their nutrition, or taking more recovery-days, they preferred to mask the symptoms with muscle strapping or compression tights, both of which I address later on. Our customers also frequently purchased expensive, lightweight running shoes in the hope that it would shave vital seconds off their 10 km personal best. But the same customers were extraordinarily reluctant to review their training programs or change their running style, even though the latter two would have had a substantially greater influence on performance times. Treating the symptoms is quick and easy; it requires little time or commitment, one can generally throw money at the problem, and it's amenable to subjective changes and perceived benefits. But treating the cause takes time, effort, an investment, and discipline, commodities that are in short supply in today's rat race. Many products become popular because they offer a shortcut without the corresponding compromise, embodying a *one quick fix* ethos. Be it a pill, a diet, a sports garment, or a training program, the claim is that the product will make you leaner, faster, stronger, and more intelligent, with minimal effort or expense. It sounds almost too good to be true, and it usually is.

The ergogenic aid also aligns with the human propensity for immediate gratification. In 1943, American psychologist Abraham Maslow proposed a model on human motivation which he called Maslow's Hierarchy of Needs, and it offered a simplified insight into the reasoning behind decision-making. The model suggested that lower-level needs (basic physiological functions like eating) must first be met before an individual will be sufficiently motivated to satisfy the so-called higher-level desires such as love, fostering a sense of belonging, self-esteem, etc. Although academically contested, the model is intuitive because it aligns with our experiences. For example, some struggle to find the motivation to read or meditate if they're feeling unsettled or frustrated at work. Similarly, none of these needs are likely to be fulfilled if the individual is hungry or cold. So, in a simplified model, basic functions take priority, and there's an evolutionary pressure to seek out simple contrivances that satisfy them. The catch is that short-term pleasures are, by definition, transient. They can be easily achieved (in most cases) but they quickly expire.

Research suggests that such instant gratification is in direct competition with long-term strategizing, although it's evident that both play a role in effective goal-setting. A collaborative study from researchers at Princeton, Harvard, and Carnegie Mellon Universities presented a simplified model of the way in which the various brain regions function in the process of decision-making. This research in the field of neuroeconomics – a study of the mental and neural processes that drive economic decisions – was published in the journal *Science* [5]. The study participants were asked to make a series of decisions that pitted small monetary rewards that were immediately obtained, against larger rewards for which they'd have to wait; it's the grown-up equivalent of the famous Marshmallow Test used in children. While the subjects made their decisions, the researchers recorded brain fMRI scans. While all decisions were associated with brain regions known to govern abstract reasoning, decisions that bestowed short-term rewards were more likely to activate neural circuits associated with emotion. David Laibson of Harvard explained:

> Our emotional brain has a hard time imagining the future, even though our logical brain clearly sees the future consequences of our current actions… Our emotional brain wants to max out the credit card, order dessert and smoke a cigarette. Our logical brain knows we should save for

retirement, go for a jog and quit smoking. To understand why we feel internally conflicted, it will help to know how myopic and forward-looking brain systems value rewards and how these systems talk to one another.

Short-term gratification is associated with a very potent chemical response which triggers dopamine-related, incentive-driven behavior, as well as positively valanced emotions. Put simply, they make us feel transiently good. Other reward systems that integrate serotonin and adrenaline (evoking feelings of wellbeing and exhilaration, respectively) also play an important role in reward. Despite the simplified explanation, these responses compete with the brain's reason centers which implore long-term adherence. Logic and reason are no match for emotion, particularly when the latter is underpinned by such an intense chemical cocktail. It supports the notion that we're compelled to favor an immediate, short-term advantage over actions that might benefit long-term success.

In the applied context, this explains the post-training candy bar we later regret, our difficulty adhering to diets, and our obsession with all manner of sporting gadgets and contrivances. A marathon training shortcut manifests as a supplement, some new trainers, an electronic gadget, all the devices that are rendered worthless unless one simply accumulates sufficient mileage and time on the feet. By contrast, the most favorable long-term strategy would be to design an effective training program, periodized over months or even years, to enact dedication and discipline, maintain consistency in sessions, eat right, sleep well, and rest when appropriate, all of which are difficult, costly, and require a substantial investment.

Finally, it's also likely that ergogenic aids align with tribalistic traits for which humans are notorious. Tribalism is the desire to belong to a group, in which we conform to a set of cultures, behaviors, or traditions. In many people, tribalism is exhibited in the staunch defense of a religious or political ideology, the support of a sports team, membership of a fraternity, or arbitrary nationalism. Humans tend to associate with like-minded individuals because we likely feel more comfortable in their presence, and develop an affinity for shared beliefs. From his book *12 Rules for Life*, Clinical Psychologist Jordan Peterson suggests:

> People who live by the same code are rendered mutually predictable to one another. They act in keeping with each other's expectations and desires. They can cooperate. A shared belief system, partly psychological, partly acted out, simplifies everyone – in their own eyes and in the eyes of others.

Even scientists and skeptics, with their strong critical faculties and independent thoughts, aren't immune to tribalistic traits. For example, we're forever indulging an obsession for science-fiction and congregating at conventions, even though organizing skeptics at a convention has been likened to herding a clowder of cats. But there's no denying that tribalism is part of our genetic makeup; it's a normal and integral component of society. Studies have characterized the hormonal responses of people exhibiting group behavior, and found that oxytocin – often referred to as the *love hormone* because of its high concentrations in the brains of new lovers – is also considered to be a *tribal hormone* because of its role in the promotion of altruism and ethnocentrism (in-group identity)[6]. Ergogenic aids feed our tribalistic habits because they act as an identifier for an athlete or exerciser within a group. Owning the latest bike, heart rate monitor, running shoes, sports garment, or magic bracelet reinforces membership of and position within the group, while distinguishing insiders from the outsiders. Owning the right bike, the latest carbon frame, or specialized pedals is often a prerequisite to high-level performance in cycling and triathlon. Those most precious about their bikes are arguably the amateurs who strive to emulate the pros, many of them paying large sums of money to save a few hundred grams on a pair of cycling shoes, even though they could often lose several kilograms by shedding excess body fat; again, it's simpler to throw money at the problem, albeit less effective. This group of athletes have also been known to ritualistically shave their bodies. It's claimed that shaving improves aerodynamics and reduces performance times, as it might in swimming. But while saving a few seconds in a race might benefit a pro-cyclist, the advantage to an amateur (beyond placebo) is likely negligible, with far greater gains available from less extreme measures.

All these products and practices do more than offer shortcuts to success or inspire us to exercise; they cement our status in an exclusive club of *those who train*. But we need to be mindful of the impact they can have on our behaviors and wider culture. As we discuss shortly, the sale of product is poorly regulated, and marketing campaigns are specifically designed to expose our scientific illiteracy, our engrained shortcut-seeking behaviors, and our psychological insecurities.

1.3. A History of Health Claims

Snake Oil and its modern derivatives have been deceiving the masses for centuries. We're not alone in the fight to resist dishonest salesmanship, but it may feel like it on occasion with the regulations often so lax. What follows is a brief review of the complex structures governing product manufacture and associated advertising claims. This short discussion may help contextualize why regulations currently appear to be unenforced, and why books like this are pertinent. The simple fact is that the governing bodies tasked with protecting consumer interests are failing; on some products, they're overly strict, and on others, they're unfathomably lenient. Plus, manufacturers and suppliers have become shrewd in their attempts to circumvent federal sanctions on advertising claims. It's up to us, therefore, to sharpen our critical faculties sufficiently to take on the burden of protecting our own interests. Clark Stanley's enterprise, though modest by contemporary standards, was a commercial powerhouse in the late 1880s. Despite high demand for his Snake Oil product, Stanley's nation-wide production facilities were eventually forced to close following an investigation (and subsequent fine) resulting from the Pure Food and Drug Act of 1906. This was the first of several laws, signed by US President Theodore Roosevelt and implemented by the Federal Government, which championed *consumer protection*. The act enabled closer scrutiny of foods and drugs that crossed the interstate, and necessitated that active ingredients aligned with the standards of various governmental science agencies including the US Bureau of Chemistry. Chiefly, however, the act moved to ban products that were mislabeled or otherwise misrepresented.

In the US, the Pure Food and Drug Act of 1906 was a milestone in consumer healthcare which shaped current regulatory structures. It forced manufacturers to state, on the product label, any of ten ingredients that had been deemed *addictive and/or dangerous*, including any that were psychoactive or sedative drugs like alcohol, morphine, opium, cannabis, or stimulants. The act later contrived to ban unsafe drugs and, eventually, outlaw products that were safe but not effective. The legislation is, perhaps, most famous for challenging the Coca-Cola brand for its excessively high caffeine content. The carbonated soft-drink, first developed by Dr John Stith Pemberton in 1886, was originally made by infusing sugar-syrup with high concentrations of cocaine (five ounces of coca leaf per gallon of syrup). Cocaine was removed from Coca-Cola in 1903 following anti-narcotics legislation, only to be replaced with caffeine. Following a legal battle, the supreme-court ruled that Coca-Cola had to reduce their caffeine content, forming the basis of the product we know today. Interestingly, the Cola now available in stores worldwide still contains coca, but with the ecgonine alkaloid extracted, at a New Jersey chemical processing facility, in order to nullify its psychoactive effects. The facility is heavily guarded.

The Food and Drug Administration (FDA). Eventually, the 1906 act garnered enough support to coalesce with other federal agencies and originate the FDA. Following many extensions and revisions to include food, drugs, and cosmetics, the modern FDA should have a strong relevance in today's exercise supplement industry, rife with products like mass gainers, fat burners, and stimulants. But although the FDA will confront products it deems to be *unsafe* or otherwise *ineffective*, the agency still approves health claims that, based on an ambiguous threshold of what's considered to be scientific evidence, show a link between a stated ingredient and a given condition or disease. Due to the restrictions, manufacturers cannot make overt statements that a supplement will cure (or treat) a disease, but only that it might *minimize the risk*. We explore the FDAs principal function later in our discussion on *Supplements and Drugs*.

There are several other watch-dog companies that are monitoring advertising standards on our behalf, regulating claims relating to dietary supplements and sports products in the wider context. In the United States, the Federal Trade Commission (FTC) is a bipartisan agency (formed of a political coalition) that functions to protect consumers. The unambiguous FTC mission statement, from their website:

> The FTC protects consumers by stopping unfair, deceptive or fraudulent practices in the marketplace. We conduct investigations, sue companies and people that violate the law, develop rules to ensure a vibrant marketplace, and educate consumers and businesses about their rights and responsibilities. We collect complaints about hundreds of issues from data security and deceptive advertising to identity theft and Do Not Call violations, and make them available to law enforcement agencies worldwide for follow-up.

There is, however, no UK equivalent of the FTC. The closest regulating body is the Advertising Standards Authority (ASA), which is an independent regulator of advertising across all media. The ASA features on its website a section entitled *non-compliant online advertisers.* In their own words, the ASA *expects all online advertising to be legal, decent, honest, and truthful*, and provides several examples in which manufacturers have broken the rules despite repeated requests for change. For example, the ASA reports a company called Protein World for claims related to their *Carb Blocker* product which weren't authorized on the EU Register. Another supplement company – Power Body Nutrition Ltd – was reported when it was claimed that their THYROID T3 supplement could *burn fat fast* and *boost thyroid output and metabolism;* these, too, were claims unauthorized on the EU register, and the ASA concluded that both adverts breached the Code. There are two major issues, however, with such regulatory action; first, despite its ongoing efforts, the ASA is thinly spread because they're tasked with pursuing tens-of-thousands of consumer complaints regarding television adverts, many of which do not pertain to sport or health. Their reach is, therefore, limited. Second, the ASA can request that manufacturers change their advertising strategies but they can't legally enforce it, and repeated requests for compliance are frequently ignored. Another UK regulating body, The Medicines and Healthcare products Regulatory Agency (MHRA) of the United Kingdom, is an agency of the Department of Health and Social Care responsible for ensuring that medicines and medical devices work to a safe and effective standard. Among their critics is Ben Goldacre who, in his book *Bad Pharma*, accused the MHRA of advancing drug company interests and prioritizing them over those of the public. The MHRA and the FDA of the United States are the two principal regulatory authorities responsible for the authorization, granting, renewal, variation, suspension, and revocation of licenses for medicines and medical devices in the UK and US, respectively, while the European Medicines Agency (EMEA) manages the process in many European countries.

A Shift of Power. While there's more regulation today than ever before, it's axiomatic that more stringent policies are needed, and we're still bombarded by products with claims largely unsubstantiated. Presently, it's salient that while there's some overlap in the policies regulating clinical *drugs* and exercise *supplements*, they're largely distinct; there is generally more scientific rigor applied in the former. In the pharmaceutical industry, drug companies are legally obliged to provide high-quality evidence of safety and efficacy of their products, and may only make overt claims in their adverts that have been first approved by the FDA. But in the dietary supplement industry, politics and continued lobbying by supplement sellers eventually led to the Dietary Supplement Health and Education Act of 1994, which effectively undermined the work of the FDA, and crippled regulations. The act liberated the supplement industry from stringently testing their products. Recent applications to amend the act and return power to the FDA have been blocked, leading to the meager authority the FDA now have with which to enforce changes to stated ingredients. Indeed, the FDA can only forcibly remove a product from sale if that product contains a regulated drug, or if the FDA can prove that the product has caused harm to the consumer. Accordingly, we are amidst a twisted regulations paradox, one in which *the burden of proof is on the FDA to empirically demonstrate that a product is dangerous, rather than on the manufacturer to empirically demonstrate that it's safe*. The ASA in the UK threatens with the same toothless bite, i.e., they make demands which they cannot legally enforce. We buy products in an antiquated system that serves manufacturers at the expense of the consumer. Often, to circumvent a lack of scientific evidence-for-efficacy, a *structure/function* claim is employed, which makes reference to a supplement's ability to maintain structures or functions within the body. Because already deemed scientifically plausible, such claims do not have to be first approved, or even reviewed, by the FDA. Instead, the manufacturers can avoid repercussions by printing a disclaimer on the product label to the effect of: *This statement has not been evaluated by the FDA. This product is not intended to diagnose, treat, cure, or prevent any disease.* The FTC has a responsibility to address post-marketing claims, but with tens-of-thousands of sports products available, each one making its own special plea to efficacy, the FDA and the FTC simply cannot keep pace. For the most part, regulatory bodies appear more concerned with ensuring that product labels accurately represent the ingredients included therein, rather than on whether the product has any claim to efficacy.

In recent years, steps have finally been taken to change the way nutrition products, specifically, are advertised, but we're far from winning the war on false claims in health and fitness. The Nutrition and Health Claims Regulation was enforced in 2012. The regulation deals specifically with supplements aimed at improving health and/or performance, and manufacturers can make overt statements only about those ingredients contained on a list of authorized claims. The new regulations were supposed to end the use of statements which weren't considered scientifically accurate in the EU. The regulator (guided by the European Food Safety Authority, EFSA) requires evidence from human studies which proves to a point of *general consensus* that there's a beneficial effect of an individual nutrient. A list of authorized health claims has been published at the European Commission website, where you can see their definitions and search for authorized claims about individual nutrients (http://ec.europa.eu/nuhclaims). Perusing the site, it's clear where certain nuanced phrases in advertising have originated; for example, the EU distinguish between foods which are *a source of protein* (at least 12% of energy value from protein) and those which are *high protein* (at least 20% of energy value from protein). But manufacturers use cunning in their advertising strategies, and invoke clever language to imply a claim without overtly stating it, thereby circumventing the regulations. It's a loophole that's very difficult to address without successively testing each-and-every claim made for every product in the market, and regulatory bodies simply don't have the firepower.

The regulations underpinning the industry, as one would expect, are convoluted and ruled by bureaucracy. Fortunately, there's a further safeguard in the search for truth; science is a process that asks important questions, invites scrutiny and criticism, and offers no concession for those who make testable hypotheses. Researchers around the world have taken it upon themselves to evaluate the effectiveness of many of the products on sale, and these scientists are rarely under any obligation to report favorable outcomes, or support the claims of manufacturers. In Chapters 6–9, we look more closely at the claims made for a series of products sold in sports nutrition, supplementation, exercise training, equipment, and alternative therapy, and explore the pertinent research domains. One research group in particular – a team at Oxford University in the UK – examined 431 claims in 104 sport product adverts and found a *worrying* lack of high-quality research to support the claims being made; they called for better studies to help inform consumers[7]. Their comprehensive study concluded that it's *virtually impossible* for the public to make informed choices about the benefits and harms of advertised sports products. Well, that's where I beg to differ.

1.4. The Post-Truth Era

It's January 2017 at a White House press-briefing. The pertinent topic is the presidential inaugural ceremony which, according to the press secretary Sean Spicer, was the *largest audience to ever witness an inauguration – period – both in person and around the globe.* Later, when the statement was criticized for its accuracy (it was objectively and demonstrably false), Spicer's comments were defended by the campaign strategist and counselor Kellyanne Conway; she said, in a press interview: *our press secretary, Sean Spicer, gave alternative facts to these claims.*

In 2016, the Oxford-English Dictionary proclaimed their word of the year to be *post-truth*, an adjective defined as *relating to or denoting circumstances in which objective facts are less influential in shaping public opinion than appeals to emotion and personal belief.* The term has seen modest use over the past decade, until the 2016 spike in popularity which was forged in the wake of two political movements in which appeals to emotion were employed to sway mass opinion: the European Union referendum (aka *Brexit*) in which the United Kingdom voted by a slight margin to leave the EU, and the United States presidential campaign. Post-truth has also received much media attention, particularly since the 2016 award. Indeed, there are several books exploring the sociocultural implications of the term, e.g., *Post-Truth Era* by Ralph Keyes, and *Weaponised Lies; how to think critically in the post-truth era* by Daniel Levitin.

The term *post-truth* evolved, primarily, because society witnessed the emergence of career politicians building their reputations on statements that weren't objectively true and, more importantly, suffering no consequence. In post-truth politics, debates are framed by appeals to emotion, with facts and figures frequently ignored, and the term was coined to capture and articulate a sentiment of frustration with the contemporary system. For the Independent (UK) newspaper in 2016, Matthew Norman wrote: *The truth has become so devalued that what was once the gold standard of political debate is a worthless currency.*

The ethos of post-truth has been pervasive in politics for many decades, but has become more commonplace in mainstream society due to the internet and, specifically, social media, where products and policy easily bypass the lax critical-thinking filters of the general user. For example, Facebook featured heavily in the news following the 2016 US presidential campaign because of the means by which the platform generates user content. When a user scrolls their newsfeed on the website or in the smartphone application, algorithms detect the stories, videos, or images on which they pause (and for how long), as well as the items on which they click. Content is, thus, generated automatically based on the user's viewing history, ensuring they see the news-stories and articles in which they're likely to be interested. It's a sophisticated system that ensures continued custom and engagement.

Another popular social media platform – Twitter – is home to countless fake, automated accounts called *bots* (an abbreviation of robots). These bot-controlled accounts generate their content automatically based on algorithms which are designed to spam the hashtags. As such, a user's search on a given topic (using a set of *hashtags* or keywords) may return content that provides a false consensus to sway public favor to one side of an argument. One point of contention is that, as sophisticated as these algorithms appear, they only recognize when an article, page, or meme is gaining traction manifesting as likes and shares, but they're not sufficiently *intelligent* to distinguish helpful content from the unhelpful, i.e., fact or fiction. Content is simply generated on the basis of preceding popularity. Consequently, an article on the latest mission to Mars or a medical breakthrough is just as likely to gain traction as anti-vaccination propaganda or religious extremism inciting violence. Recent history demonstrates that officiating bodies do a paltry job of regulating/restricting harmful content. Such an order is, ultimately, the product of a system which values *likes* and *shares* above moral and ethical values. In a socio-cultural and physiological way, *likes* and *shares* are the drugs of our time. Learn more on the spread of malign narratives across social networks by reading the work of Dr Renée DiResta.

The result of all this is a synthetically generated echo-chamber in which consumers are not exposed to both sides of a given story, only the side that conforms with their pre-existing notions. This is decidedly problematic because so many individuals, particularly young people, obtain their news and insights into world events via social media. Such rampant use of these platforms makes it easy for post-truth to bleed into every facet of our culture. A climate in which objective fact – underpinned by scientific research and evidence – can be systematically ignored in favor of emotional language and bravado breeds grand and fallacious claims that go largely untested; and it's just as much the theme of contemporary politics as it is the lynchpin of health and fitness marketing. This overt propagation of misinformation and falsehood has rendered independent critical-thinking an essential prerequisite to a healthy perspective on the world.

One of the great hallmarks of our time is that, in developed societies, we're encouraged to offer our opinions on any number of divisive issues, emancipated from oppressive acts of retaliation, censorship, or sanction, assuming we're not slandering or inciting violence. Freedom of speech is true to the ethos of both democracy and scientific enquiry in that nothing and no one should be above scrutiny, but it comes at a cost. The post-truth culture has society questioning the validity of self-evident facts. For example, we coexist with fringe activists who feel confident in contesting the fact that the Earth is round. Indeed, the so-called Flat-Earthers assert that the world is flat, and that a global conspiracy led principally by NASA is deceiving the masses with false documents and feigned satellite footage. Why it has to be NASA, and exactly what the masses are being fooled into, is anyone's guess. If conservative estimates that 1% of the US population are *flat-earthers*, their number equate to more than 3 million. Such groups are afforded platforms in the guise of pervasive social media in which everyone is an expert, and all opinions are of equal value. For this notion, there are two considerations: first, an opinion on a specialist topic is only valid if espoused from someone educated on that topic; for example, I'm in no way justified in commenting on the hydrodynamics of shark fins or the amount of fuel required to blast a rocket into geocentric orbit. In such respects, my opinions simply aren't valid. Second, everyone is entitled to their own opinion, but everyone is *not* entitled to their own facts. Facts, by definition, are objective, can usually be quantified, and have stood rigidly in the face of repeated tests. Opinions are subjective, personal views/judgments on a topic, open to bias, and that aren't necessarily proven. The Earth is objectively and quantifiably round (it's actually an oblate spheroid), and there are numerous calculations, images from Earth satellites, and expert agreement on this documented fact. Yet, it's one still contested. But there's no use in arguing facts in a climate where facts aren't regarded; you cannot reason somebody out of a position they didn't first reason themselves into. But while the consequences of a flat-earth rhetoric are relatively minor (a small proportion of the population believing emphatically that the Earth is flat won't in itself infringe the development of modern medicine, obstruct space exploration, or influence climate change policy), there may be far broader implications of such a growing culture of anti-intellectualism, and the false equivalence of objective fact and subjective opinion.

With respect to health and fitness, the industry is predicated on advertising, marketing, and grand claims to efficacy, all of which thrive in a *post-truth* society. Fortunately, there's a growing movement of passionate scientists and skeptics rallying to contest the notion of post-truth that's threatening the ethical fabric of our society. Indeed, many organizations have been established to empower consumers to protect themselves and each other against false claims. From www.truthinadvertising.org:

> Each year, American consumers lose billions of dollars as a result of deceptive marketing and false ads. These run the gamut from blatant lies and fraudulent scams to subtle ploys intended to confuse and mislead. Not only do these tactics impact us as consumers, but a mind-boggling amount of money is misdirected in our economy as a result of deceptive marketing

Not all advertising and marketing is flawed or misleading; an intelligent, measured, and well-considered advert can be entertaining, inspiring, and can offer an insight into quality products congruent with claims that are based on evidence. And circumstances have improved in some respects; there are now rules and regulations to which companies must adhere, and examples of legal consequences to misleading and false advertising. For example, in 2013, the National Advertising Division (NAD) took issue with a claim made by NITRAMIX regarding a caffeine-containing product they claimed was *clinically proven to increase strength*. Following their investigation, NAD concluded that *the advertiser relied in part on a pilot study that included only five participants and lacked appropriate blinding and placebo controls*. This is a sound judgment from the NAD, exhibiting that they have a reasonable standard for what can be considered 'clinically-proven', and NITRAMIX didn't meet it. For a second example, we look to the MusclePharm Arnold Schwarzenegger series. What better way to advertise a strength-training supplement than to align it with the world's most well-known body builder, seven-time Mr Olympia, movie star, and ex-Governor of California. The class-action lawsuit filed in 2015 alleged that MusclePharm misrepresented the amount of protein in its *Arnold Schwarzenegger Iron Mass whey protein* product, and that the supplement contained about 50% less protein than that indicated on the label. The company was also accused of engaged in *protein-spiking*, a process by which the product's nitrogen content was artificially increased using non-protein-containing ingredients that are cheaper to produce. A third and final example is a class-action lawsuit filed against NBTY and United States Nutrition toward the end of 2017 for falsely advertising *Body Fortress 100% Pure Glutamine Powder*, alleging that the marketing of their supplement as the *ultimate recovery fuel that boosts post-workout recovery* was misleading. On closer scrutiny, there was insufficient evidence to support the claims (vague as they were), and the lawsuit at the time of writing is ongoing (*DeBernardis et al v. NBTY, Inc. and United States Nutrition, Inc.*, Case No. 17-cv-6125, N. D. IL.).

These three were selected from a multitude of legal action suits against product manufacturers that appeared in a single domain of the health and fitness industry. While the close scrutiny of such claims, and the consequences wrought, is encouraging, the sheer magnitude of the task results in many (indeed most) claims going untested, particularly those outside the scope of sports supplements that aren't being ingested. Greater ownership over our decisions, by upskilling our ability to think critically in the face of overwhelming odds, is an essential prerequisite to successfully navigating this post-truth era with our health, our pride, and our bank balance intact.

1.5. Failures in Education

Unfortunately, the educational systems of which we're a product have rendered us unqualified to comprehensively navigate such a commercialist world, and this is particularly true in Great Britain and in the United States. Throughout my childhood schooling, two hours per week were devoted to the study of Religious Education (RE) – a subject that didn't contribute to your final grade – and a token one hour on Physical Education (PE) which, by all accounts, fell far short of the World Health Organization's physical activity recommendations. There were no classes, however, on logic, critical-thinking, or the merits of rationalized argument. In recent years, critical-thinking has begun to replace General Studies, but the majority of universities do not offer merit for the classes. In the US, most teachers don't train their students in critical-thinking for the simple reason that there's insufficient time. State education departments decree that so much material must be covered in the short academic term that non-essential content like critical-thinking is among the first to be omitted. Rarely is it even considered among the after-school extracurricular activities. On why this may be the case, Ben Morse at the Guardian newspaper (UK) suggests:

> Schools are businesses (now so more than ever, as they balance their books one playing field at a time) and the bottom line is league tables. If not nationally, then locally. A* to C pass rates that can be printed in 72-point text in promotional literature, attracting parents, their darling children, and with it the pupil premium. If a school's primary purpose is to generate pass rates and percentages into higher education (and it currently is, like it or not), then why devote staff to critical thinking? Let it fill up part-time staff's timetables, or pad out that under scheduled music teacher's day. Motivated students question what the point is to the subject, and while I would love to explain to them its long term importance, at 17, with a stack of English essays and biology research to write up, all of which has a concrete pay-off in a year's time, even they struggle to see the benefits.

In four independent UK universities at which I've lectured sports science/kinesiology, critical-thinking hasn't been a staple of the curriculum. To upskill my students, I've designed and implemented short courses in which I introduce them to the founders of rational thought and some of the pitfalls of flawed argumentation. We discuss the logical fallacies (Chapter 3), how to structure an argument, superstitious thinking, and specific examples from sports science where bold claims collapse under close scrutiny; it's well-received and proven to be essential study for anyone training for a future career in the sciences.

Since its inception as a means of communication among a niche groups of physicists at CERN, the internet has expanded at an exponential rate. We no longer have at our disposal a few trusted sources of information, but endless libraries of data and statistics and news reports, none of which are required to be independently substantiated when published online as an opinion piece or blog post. It's for this reason that I discourage my students from referencing websites (except for official governmental sites like the WHO, for example) to substantiate their factual statements. Our use of the internet and social media has grown at a rate that dramatically outstrips the capacity of our critical faculties to cope with the extra data. Indeed, if there's one domain in which we're indulged, it's in our access to new and emerging research and information. We have, literally at our fingertips, an endless supply of data which can be accessed at any time of the day or night and, in most societies, such data is freely available. But we have not yet evolved the inherent tools to decisively distinguish information from misinformation, and information is not knowledge. As depicted in Figure 1.2, information is represented by numerous individual data points, largely distinct from one-another. It's only when one knows how to interpret the data, how the data points link together,

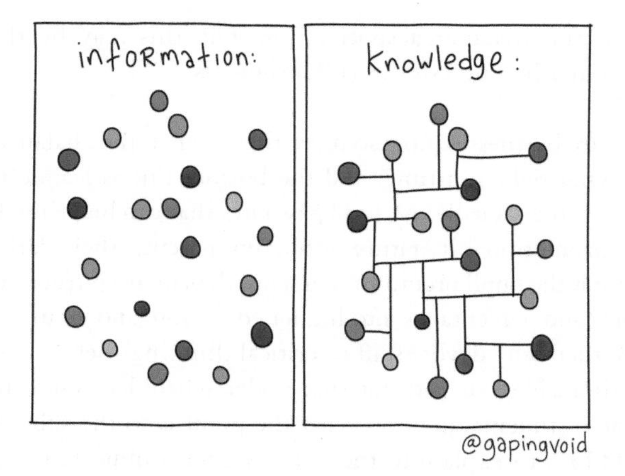

FIGURE 1.2 Information versus knowledge.

and how the data fit into the broader picture that one obtains true knowledge (Figure 1.2); this is not something that can be learned from a simple Google search. More importantly, rather than bombarding students with ever more information, relatively more emphasis should be placed on equipping them and consumers alike with the tools that will enable them to distinguish for themselves the difference between information, misinformation, and knowledge. By this I mean critical-thinking, logic, critical appraisal, and the nature of evidence. We must contrive a means of shifting the focus away from brief, one-stop online searches, and instead emphasize the importance of collating meaningful data (plural), filtering it for validity, and synthesizing coherent and evidence-based conclusions.

By way of demonstration, I typed into a search engine the following: *improve my sports performance*. The first 27 responses all linked to company websites and blog posts, none of which are obligated to provide scientifically credible advice on my search topic. That's not to say the advice was necessarily false or misleading, but I had to skim past the first 27 results before there appeared any links to published manuscripts or an education institution. I also conducted a search for *weight-loss drugs* which returned links to legitimate outlets like the MAYO Clinic and the National Health Service (NHS) interspersed with those to non-scientific weight-loss magazines and women's health forums. It may be that use of the keyword *drug* provoked a stronger academic return. Nevertheless, how is the average non-scientist without formal education in this topic supposed to know which outlet is going to provide the soundest advice? Moreover, if one is conducting a search because they're hopeful that a weight-loss drug might be the magical elixir to remedy their weight issues, they're more likely to favor sensationalist news outlets which proclaim miraculous outcomes, over conservative clinical studies with their tempered conclusions and risk-to-benefit assessments.

Earlier, I alluded that available time might be one reason why critical-thinking is excluded from formal curricula. Another is that school teachers may be preoccupied with preparing students for the practical world of work, that is, bestowing upon them the skills to learn a trade like information technology, carpentry, applied science, mathematics, plumbing, business, engineering, or marketing. While it's true that there are few instances in which critical-thinking is employed as an actual trade, its tenets underpin every facet of contemporary society, including our political system, science, research, and healthcare. There's no instance in the modern world where critical appraisal isn't considered an essential skill; inspect the person specification of a job advert (in almost any industry) and you'll no doubt see critical appraisal or an equivalent skill in the desirables column. How do we decide if researching a new medicine is worth a large government investment? How can we know if a rocket will produce enough energy to propel a space shuttle to escape velocity and leave the Earth's atmosphere? How can we discern if a political party is dealing transparently in their policies, or instead misleading the public with false statistics and an emotive rhetoric? How can we predict if a business deal is likely to prove financially rewarding? The answers to all the aforementioned are elucidated by making testable hypotheses, taking measurements and making observations, critically discussing the evidence, and arriving at valid conclusions; it's *critical-thinking 101.*

An exercise I've successfully implemented with my university students, to reinforce the importance of broad critical-thinking skills, is the *new laptop challenge*. I ask the group to recall the last time they made a significant financial investment like buying a laptop, or a car, a house, a new television, booking a holiday, or buying a mobile phone. I ask them to consider what processes or strategies they followed in coming to a decision on make, model, manufacturer, and price. For example, did they go straight to the computer store and ask the assistant to recommend a device, and then accept the suggestion on blind faith? Scarcely is this how we arrive at most decisions. Most would admit it'd be credulous to take verbatim the advice of the salesman. Did they speak to their friends/colleagues to see which models were favored and which ones they'd recommend? Did they compose a shortlist of devices that were appropriately priced, and then list the pros and cons for each one by way of direct comparison? Did they read online reviews of the product to discern a user-consensus, and were their opinions swayed by what they read? Did they conduct any research on performance issues with each device? If the item in question was substantially priced, and its functionality of relative importance, it's more than likely that the buying process would have involved a holistic combination of all of the above, in varying degrees of magnitude. In most cases, the student admits to collecting *as much data as possible*, from *a number of different sources*, and balancing the information to make *an educated, objective decision* on which product was most appropriate. Rarely did they accept the opinion of a single individual and even more rarely was the decision-making process hurried. Pertinently, they did not assume that the claims made by any given manufacturer were necessarily true, and they recognized that there was no reason to accept one manufacturer's claim over another. In general, most students demonstrate an innate propensity for critical-thinking and objective decision-making, even if these skills do not manifest in all other aspects. These are traits that should be nurtured, honed, and sharpened as a fundamental component of education, and not relegated to an afterthought to be addressed after graduation.

Accordingly, although this is a book principally concerned with the health and fitness industry, I also hope to address some of the more fundamental questions about the nature of logic and problem-solving in daily life, because these are essential tools to living confidently in today's society. A recent commentary in the New Scientist suggested that:

> Science and technology shape our world and, as a society, we need to make well-reasoned and scientifically literate choices about everything from genetic engineering to geoengineering.

With respect to education in the sciences, it'd be wrong to assume that it's only lay-people who are subject to bias and flawed thinking; scientists, too, have often been denied a direct education in critical appraisal, frequently developing these skills throughout the course of their careers. Sometimes, however, if you shift a scientist or researcher minimally outside of their esoteric area of study, their critical faculties shake and crumble like a house of cards in a soft breeze. This is when you happen upon medical doctors prescribing homeopathy (a discredited and highly unscientific alternative medicine), astronomers who believe in astrology (although this is much rarer), and nutritionists espousing the benefits of superfoods and detoxing. In 2012, I received an email via my blog from a scientist with a Ph.D. in quantum mechanics; this is the branch of physics concerned with the behavior of fundamental particles. She staunchly believed that the Moon's orbit could influence human behavior by exerting a *pull* on body water, in the same way it exerts a gravitational pull on the ocean, causing tides. The adult human body is around 70% water; the plumpness exhibited by newborn babies results from a greater water content (up to 90%), while the sagging skin of an old-age pensioner is attributable to a considerably lower water content. Given the predominance of body water, it's understandable how the non-scientist might see the claim about the moon as superficially plausible, but it nevertheless requires just some rudimentary study to render the notion false. The gravitational pull of the moon has been calculated precisely and influences humans to an equivalent extent as a mosquito as it sucks the blood from your arm (i.e., negligibly). Yes, the moon is substantially larger than a mosquito, but it's also considerably further away (about 240,000 miles further). The claim is demonstrably and empirically false. While acknowledging that quantum mechanics and astro-physics are distinct specialties, the laws of gravitation are well understood by all branches of physics. Moreover, of all people to have assimilated the relevant mathematical formulas of gravitation, one would have expected a particle physicist to have been among the first; yet, on this occasion, she exhibited a profound ignorance of some basic scientific principles, and favored an unscientific, widely discredited hypothesis.

More recently, a colleague of mine – a cancer biologist – expressed his disbelief in the theory of Evolution by means of natural selection, on the basis that it's *just a theory*. This isn't a book about Evolution (or astronomy), and so I don't intend on expunging any creationist arguments here. In short, scientists are as confident in the validity and explanatory power of Evolution as they can possibly be about any natural phenomena. There are libraries of books elucidating the reams of direct and indirect evidence in support, available for anyone seeking to expand their understanding. My colleague's objection to Evolution was not on the grounds of evidence, but rather that it contradicted his preconceived ideology. In defending his position, he perpetrated a deliberate misuse of the term scientific *theory*, which is the highest accolade bestowed upon any well-studied and supported principle. A scientific *theory* is a hypothesis (an educated guess) which has, in turn, been reinforced by multiple independent lines of evidence. The salient point is that for a cancer biologist to deny Evolution requires the most sophisticated contradictory viewpoints (cognitive dissonance), a denial of evidence, and a profound ideological bias. We're all subject to making mistakes in logic and reason, and formal education in the sciences isn't necessarily an effective inoculation.

1.6. Carbohydrates, Vaccinations, and the Pope

The earlier discussion of *post-truth* provides a backdrop to further explore pitfalls of ideology within sports science, to which nobody is immune. The Flat-Earth movement is an ideology (a system of ideas and ideals) that isn't based on any objective fact and which can, for the most part, be dismissed as a quirky and relatively benign product of the current *post-truth* climate. When there's evidence to support both sides of an argument, however, opposing positions will be vehemently defended based on similarly strident ideals. An ongoing debate within health and nutrition is the relative benefits and contraindications of a low-carbohydrate versus a low-fat diet. Proponents of the former argue that refined sugars are the mechanistic underpinning of many of the world's clinical ailments on the basis that the original research on the causative link between dietary fat and poor health was scientifically flawed. There's insufficient scope here to do justice to the research, nor accurately represent the respective position stands which have become complicated and convoluted. For the purpose of our theme, I'd instead like to highlight how both standpoints are often vehemently defended based, at least in part, on ideological stances.

On both sides of the above-mentioned debate, there are world-leading experts – medical doctors and career scientists – engaging in heated and often vitriolic argument. Criticizing one position renders you vulnerable to personal attack and abuse from the other; if my words hold no sway, scrutinize some of the relevant Twitter feeds, or the scathing letters published in academic journals. These are fascinating discussions, played out constructively for the most part, but at times becoming petty and unhelpful, and devoid of logical discourse, rather like the UK House of Commons. Moreover, both sides are capable of citing reams of published literature and expert statements to support their position and desecrate the other. And yet, critical to our present assessment is that one side is wrong, objectively and decisively wrong. Which one of them is mistaken is unclear, and more dedicated research needs to be undertaken to nuance the debate. The point here is that aligning oneself to a given camp, even if based on an interpretation of the evidence, has the potential in some to generate a tribal bias and, by definition, tribes have a tendency to oppose evidence that is contrary to their position. Once you've chosen sides in such a divisive debate, you have inherited a form of bias and it's very difficult to be authentically objective. Indeed, if the evidence supporting your side's argument is proven false, then you, by virtue of your position, will also be proven false; and nobody likes to be false. When objectivity is compromised, and people make decisions based on pre-existing beliefs and ideology, then logical discourse crumbles. Taking an ideological stance will unquestioningly hamper the ability to get to the bottom of important questions like this, and thus establish fact. I willingly defer on the debate because I've not read and assimilated all of the available studies. Nevertheless, I'm educated on a much greater proportion of the research than some of those who've already chosen sides in this heated dispute.

Bias is a dangerous thing because it affords motive to disregard facts and evidence that to others are as clear as day. Strategies to eliminate bias should be the first lesson we teach children at school; we discuss bias in more detail in *Chapter 5: Placebo Products and The Power of Perception.* In this chapter, I've provided real examples of how an astrophysicist, a medical doctor, and a post-doctoral research scientist exhibited flawed thinking which led to incorrect conclusions about the natural world. Even if you're without the foundations of a science education, I firmly believe that by assimilating the arguments in this book, you'll be in a stronger position to critically appraise assertions than many classically trained scientists. Indeed, the smartest, most logical, rational, and self-aware individual I know doesn't own a science degree, or any degree, but has obtained and refined their outlook through years of independent study and rational debate. You already have the raw tools; they just need sharpening. And this will be our task as we proceed on our journey to expose the myths and fallacies of sports science.

2

SHARPEN YOUR TOOLS

a·ware·ness.

Concern about, and well-informed interest in, a particular situation or development.

2.1. Consciousness-Raising

In the last chapter, I laid some foundations and contextualized the re-emergence of the Snake Oil salesman by discussing post-truth, critical-thinking education, and the history of health claims. I also introduced some of the regulatory bodies responsible for protecting consumer interests in the health and fitness industry, where under-developed critical-faculties coupled with poor regulations leave manufacturers unaccountable for their exaggerated claims. Next, we focus on the tenets of critical-thinking with specific emphasis on how logic and reason can be twisted and fractured in the marketing of health and fitness products. To emphasize the scope of the issue, we'll now focus on the immediate task of *consciousness-raising*. In his 2006 book *The God Delusion*, Richard Dawkins wrote:

> In a science-fiction starship, the astronauts were homesick: 'Just to think that it's springtime back on Earth!' You may not immediately see what's wrong with this, so deeply engrained is the unconscious northern hemisphere chauvinism in those of us who live there, and even some who don't. 'Unconscious' is exactly right. It is for a deeper reason than gimmicky fun that, in Australia and New Zealand, you can buy maps of the world with the South Pole on top. What splendid consciousness-raisers those maps would be, pinned to the walls of our northern hemisphere classrooms.

> Day after day, the children would be reminded that 'north' is an arbitrary polarity which has no monopoly on 'up'. The map would intrigue them as well as raise their consciousness.

Consciousness-raising is, therefore, not only a means of recognizing an inherent chauvinism or bias, but also a means of learning by establishing systems of thought that aren't constrained by such bias. In the present context, bias comes from our innate propensity for heuristic shortcuts, the thought processes that compel us to seek out a prevailing path of least resistance, an expedited path to the ergogenic aid. With respect to the products we buy (and the premises we buy into), it's easy to make decisions without engaging our critical-faculties, precisely how product manufacturers prefer it. We're conditioned, after all, for survival: to eat food when it's available, to spend money in the short-term rather than save for the long, and to exploit opportunities that bestow a quick, convenient path to success. In other words, we're conditioned to live as though survival wasn't assured and contingency-planning for the long-term wasn't a priority. Douglas Kenrick and Vladas Griskevicius, from their book *The Rational Animal*, suggest: *What seems foolish and even delusional from a traditional perspective can be smart from an evolutionary vantage point.* But what does this mean for humans who now live in a world oriented more around work and leisure than basic, every-day survival?

Presently, manufacturers are experts at understanding and exploiting behavioral psychology to influence our buying decisions. Indeed, most large retailers employ psychologists for this purpose. In an online article for the Brooklyn Fashion Accelerator, Laura Moffat discussed the reasons behind our dramatic increase in buying-habits:

> [Fast fashion]...cheap, low-quality, trend based items that are here today and gone tomorrow. Picked up from the runway and re-engineered for the everyday consumer in a matter of weeks. We go wild for this type of fashion. Like crazy-mad-can't control ourselves from buying MORE followed by self-justification of why we needed that exact same t-shirt we bought last year. We used to value longevity and craftsmanship and where our garments were made. Now we are in a fast-paced rat race that is filling our closets with unnecessary and often unworn clothes. In fact we only wear 20% of the clothes in our closets, and as a nation, Americans throw away 11 million tons of textiles per year! [sic]

We must be careful not to prefer health and fitness the same way as our fashion, entertainment, social media, and our food, i.e., quick, easy, accessible, and affordable. The associated chemical responses orchestrated by our primal, ancestral brains have evolved to make these behaviors pleasant and rewarding in the short-term, while offering little in the way of long-term value. An awareness and acknowledgment of this is an important step in retraining a mind to think more critically, so that engaging logical, rational faculties will eventually become an automated response. Below, I present an example from mainstream culture that exemplifies how our primal biases can be exploited to influence our decisions, and a second example that develops the general theme of bias. I finish the chapter with an absurd thought-experiment I'd like you to attempt in an effort to raise our collective consciousness.

2.2. Dihydrogen Monoxide

I have a love/hate relationship with the internet; this most profound invention has forever altered the means by which we work, play, and communicate. But perhaps its greatest progeny to date is the internet meme. Specifically, the meme associated with parodying dihydrogen monoxide (DHMO). The chemical first received widespread coverage in 1993 when a Michigan newspaper reported how a vast quantity of DHMO had been discovered in water pipes, that it was fatal if inhaled, and that a vapor of the chemical was capable of blistering the skin. There was a renewed concern about DHMO in 1997 when a 14-year-old student named Nathan Zohner began a petition to have DHMO banned due to its potentially harmful effects. Internet memes from the last decade have emphasized some alarming side-effects, for example: (1) *if DHMO can rust metal pipes, imagine what it's doing to your insides;* (2) *once DHMO comes into contact with your skin, it CANNOT be washed off;* (3) *DHMO is also known as hydroxyl acid, and it's a major component of acid rain;* and, (4) *they put DHMO in your milk, and also in your bleach!*

If you weren't already aware, dihydrogen monoxide is the chemical name for water (comprising two hydrogen and one oxygen atoms; H_2O). Water is important to humans; our bodies are predominantly water (somewhere between 70% and 90% depending on your age and hydration status) and for this reason, there are myths and misinformation regarding what water is, and is not. For example, water isn't inherently *good* or *healthy*, and it has no agency of its own. Some suggest that drinking between 3 and 4 liters of water per day is necessary to stay healthy and ensure normal metabolic function. But why should this be the case; if you're *adequately hydrated*, then you're *adequately hydrated*. Anecdote and hearsay propagate the notion that water helps to detoxify the liver and flush out the kidneys; in fact, water is handed out after massage on instruction from the governing body for soft-tissue therapy, but this is not an evidence-based claim. It's certainly possible to overhydrate if water is consumed with reckless abandon, and there are numerous instances of water intoxication in athletes and exercisers during marathon running, who have taken a little too literally advice from trainers and coaches to *drink plenty of water.* In such cases, the result is a dilution of body sodium, thereby decreasing its concentration, resulting in an increased risk of dilutional hyponatremia (a potentially fatal condition).

The advice should be tempered to *drink enough water to remain adequately hydrated*, and the precise implication of this for each individual should be carefully considered.

The trope of dihydrogen monoxide is pertinent because an abundance of health and fitness products are marketed on the premise that natural products are inherently *good* and unnatural (or synthetic) products are inherently *bad*. This is known as the fallacious *appeal to nature*, and it's a common theme in the health and fitness industry. The DHMO parody is a humorous but astute means of highlighting such fallacious logic, and raising our consciousness with respect to our bias against scary-sounding chemicals. Even the noun *chemical* is associated with synthetic, fake, harmful, and unnatural products, and used pervasively in advertising. Spaces on billboards and buses are monopolized by adverts for food and cosmetics *free from chemicals* which is an astonishing feat given that a pure chemical compound is just a substance that's composed of a particular set of molecules or ions. Accordingly, literally *everything* contains chemicals in various forms. The term has, nevertheless, been hijacked by an industry intent on demonizing synthetic products in favor of more expensive natural ones. The DHMO meme was propagated by those individuals who inadvertently proliferated misinformation and falsehoods about chemicals. The parody received moderate success in calling for the *banning of water*, and illustrates how scientific illiteracy and bias against the unnatural can be exploited to generate misplaced fear in something so ubiquitous, and essential to life.

2.3. Supermarket Scam?

Before we delve deeper into the types of products ubiquitous with the health and fitness industry, here are some other means by which marketing might tap into our unconscious biases to influence buying decisions. Regardless of age, sex, income, social status, or education, we all have extensive first-hand experience in supermarket grocery shopping. According to a survey by King Retail Solutions (KRS) and the University of Arizona, 77% of US shoppers buy their groceries from the so-called *big-box* non-grocers, i.e., large multinational stores that stock vast quantities of branded produce. In turn, these stores invest billions each year in understanding the internal mechanisms of our minds, and use a plethora of sophisticated techniques to manipulate decision-making. They often rely on the exploitation of engrained bias, and it begins the second you step through the sliding double-doors. One method of manipulation is to stimulate body senses, not just sights, but also sounds and smells that provoke your primal impulses. For example, have you ever gone to buy groceries and noticed the mouth-watering smell of freshly baked bread filtering throughout the store? These smells become tantalizingly intrusive, inexplicably pervading every inch of the store when most of the baked products are packaged in plastic containers or housed behind glass counters. It's difficult to resist, particularly when you're hungry, because in this state, you're more suggestible. It's precisely for this reason that we try to avoid shopping for groceries on an empty stomach, because doing so means we're more likely to make decisions with our stomachs rather than our heads. Not infrequently, the store lacks any baking facilities, and the odor is circulated using specially designed sprays or capsules that distribute vapors into heating or air-conditioning vents. There are several companies specializing in the sale of bread sprays and canisters.

Supermarkets are just as interested in *how* you buy as they are in what you buy. This is a science in itself, underpinned by just as much data and statistical analysis as most research in the natural sciences. The larger corporations invest huge capital in sophisticated market-research, the so-called *eye-tracking* studies. In these carefully controlled experiments, prospective customers are invited to participate in a mock shopping experience while wearing goggles which track eye position, gaze direction, the sequence of eye movements, and visual adaptations, with the aim of capturing habitual and subconscious behaviors. The valuable data gleamed from such studies are then interpreted by behavioral/experimental psychologists who inform the marketing professionals.

From such insights, companies can more effectively optimize product order and spacing, as well as optimize the colors and patterns most likely to attract your gaze. Many larger stores stock at least three brands of a given product. First, there's a more expensive superior brand positioned on the top shelf. Second is the value brand which is cheaply priced and situated on the bottom shelf next to the floor. Finally, there's the logical compromise between value and quality referred to as a tertiary brand. Superstores know that your selective extravagance rarely extends to the premier brand but they'd prefer you didn't buy economy; hence, these are poorly packaged with ugly designs and placed inconveniently. The tertiary brands are priced somewhere in the middle, situated at eye-level so that they're the first items you see on perusing the isle, and seem to offer a well-rounded compromise between extravagance and economy, like Goldilocks' porridge (not too hot, not too cold, but just right). Tertiary brands comprise the majority of product sales. Incidentally, products aimed at children are usually positioned a few shelves lower, aligning with the eye-level of the typical infant, or those sitting in the cart. To further manipulate buying decisions, stores will engage in *product anchoring* whereby a given product, usually one of superior quality, is deliberately overpriced to make a tertiary brand seem more appealing, irrespective of whether it's the best value-for-money.

Another more deceptive strategy in widespread use is in chronically conditioning the customer to an expected outcome. Such conditioning plays an integral role in the placebo effect, discussed later in the context of sports performance. Years ago, supermarkets introduced bright red and/or yellow labels to denote products on special offer or otherwise reduced (due to damage or imminent expiry). Over time, we've come to associate these bright demarcations with products that offer better value-for-money. Presently, stores will affix a red or yellow label to whatever they want to sell, usually on a rotation. Irrespective of the offer, the consumer infers a bargain. Years of conditioning – like Pavlov's dogs – we give preferential attention to labeled items.

As we're soon to discuss in the context of the health and fitness industry, some tactics are designed to tap into our chauvinisms regarding natural produce. Red meat, for example, is often sold in black plastic since the dark shade contrasts with the meat giving a richer, redder, and more natural appearance. Furthermore, shops have been known to sell their vegetables and fruit in earth-colored paper bags, or in large grocery baskets and hampers, e.g., containing apples, wrapped in paper to confer an impression of authenticity. Fish and cheese are sometimes sold at deli-counters manned by authentically dressed fishmongers who handle items that are similar to those in the isles, except the former have been left idle and exposed to oxygen for many hours, rather than being packaged at source. The entire organic food industry, in addition to the anti-GMO movement, is founded on a similar penchant for natural produce.

At the last is the most intriguing aspect of this industry. For many years, most major supermarkets have honored the loyalty card system, small credit-card-sized barcodes which, when scanned at the checkout, add points to your account which you trade for discounts and special offers, thus ensuring loyalty to your local store. But these loyalty schemes serve a greater purpose. Each time your card is scanned at the checkout, information on your buying preferences is filed on a central server, and added to a colossal database from which statistics are generated on everything from the most popular products for a given store, to the most frequently purchased items in a given demographic, sex, age range, geographical location, etc. Many companies have sophisticated analysis software, working with decades of data on buying histories, most likely orchestrated by statisticians doing the heavy lifting. They can discern, for example, the precise ambient temperature at which customers start favoring ice lollies over ice creams which, in turn, informs their stock control. Companies know the earliest point at which families begin hoarding for Easter or Thanksgiving or Halloween, and the produce they prefer. All of these data increase profitability, and depend on innate and predictable consumer behaviors.

The high-street supermarket is a neat example because it's familiar and also because it's evidently a highly sophisticated marketing machine. It's informed by ingenious data collection strategies, data analysis, and statistical output, all combined with an academic understanding of behavioral psychology and how best to manipulate and exploit the fallibility of the human decision-making process. The overarching lesson isn't to become cynical of supermarket chains and their marketing strategies, nor is this a book about grocery shopping. More pertinently, all marketing companies, particularly those oriented around the sale of sports products, want to manipulate decisions by exploiting engrained behaviors and ignorance of the marketing process. Much of their business depends on us making decisions without *thinking too much*, or without recognizing when flawed logic or false claims are clouding our judgment, bypassing our critical-faculties. Exercises in consciousness-raising, like this, serve to invigorate critical-faculties, bringing them to the fore.

2.4. Out of Control

To close this chapter, I'd like to retrain our attention on sports products, and the painfully meager regulations that underpin their sale. In Chapter 1, I alluded that if there's any indication that a product will sell – whether it's a pill, a supplement, a watch, a shoe, a lotion, a diet, a training program, a pair of performance pants, a magic bracelet with an invisible force field – somebody will sell it and people will buy it. As a means to test this hypothesis, I retreated to a quiet room one afternoon and contrived to design three distinct products that were superficially plausible, but ultimately ill-conceived and unlikely to be effective. As became apparent, no matter how far-fetched the idea or how tenuous the science underpinning it, an equivalent product was already on-sale. Note: the following paragraphs contain references to real physiological mechanisms, but misused and misappropriated to deliberately mislead.

2.4.1. The Fat-Burn Shirt

The premise here is simple; this is a long-sleeved compression shirt or whole-body garment that increases calorie expenditure during exercise. It's an aid to weight-loss because one burns more calories in a given training session, but it also promotes time-efficiency because one can attain the same calorie expenditure in a shorter training session. There are three potential mechanisms on which this garment could be marketed, all sounding sufficiently plausible to slip through the net. First, the garment provides vascular compression, like a typical compression garment, which increases blood flow around the body and, in turn, increases the delivery of fatty acids to muscles during exercise. These fats are then oxidized by muscles at a faster rate than during exercise with traditional clothing. Second, the compression stiffens the joints, thereby creating a degree of resistance against which the muscles have to work. Greater energy expenditure is required to move the muscles when compared to normal clothing. In a given time-period, your calorie expenditure would be augmented. Third, the garment is fashioned with a special layer of thermal-insulation that attenuates heat loss through the skin during exercise, thereby triggering an increase in core temperature. Increased core temperature elevates metabolic rate and, therefore, calorie expenditure when compared to exercise performed at the same intensity but without the garment. This latter mechanism would be pertinent during both exercise and at rest, so one could benefit by sleeping or working in the shirt. In fact, why exercise at all when you can sweat your calories away?

The market for my Fat-Burn Shirt is substantial. We're in the midst of a global obesity epidemic which is projected in the coming years to afflict one in two adults. Despite an increase in education on the importance of exercise and healthy-eating, most people still fail to meet basic guidelines for physical activity (modestly stated to be at least 30 minutes of vigorous activity, per day, five days of the week). Augmenting calorie expenditure during exercise would promote a more time-efficient workout, and greater adherence to long-term exercise regimens, in addition to increased fat-burning at rest in those who don't have time to exercise.

The physiological mechanisms described scarcely stand up to scrutiny. After the brainstorming session, I conducted an online search revealing two products already in shops that make similar claims to efficacy. The first is called the *ShaToBu – Calorie Burning High Waist Shaping Tights* which use seamless resistance bands to make muscles *work a bit harder during natural movements, like climbing stairs or walking*, leading to *muscle toning* and *increased caloric burn*. A press-release on the product (which has the appearance of a full-color advert) references a supporting study but, on further inspection, it was cited alongside the caveat that *[the study] was presented at a medical conference. The findings should be considered preliminary as they have not yet undergone the "peer review" process, in which outside experts scrutinize the data prior to publication in a medical journal.* I searched the PubMed database using key terms including 'ShaToBu', but found only links to gene therapies with obscure letter/number denotations.

The second product I found, which cannot be named for legal reasons, made similar claims regarding an increased exercise core temperature. The website stated that their body suit could increase the net calorie expenditure by up to 100 calories per hour during exercise, which isn't a minor value. Notwithstanding my incredulity on the preciseness of the numbers (there were no supporting data to corroborate the claim), if the primary mechanism pertains to an increase in core temperature, then a similar response would be attainable by simply wearing more clothes. You could also train indoors, turn off the fan, or turn up the thermostat. Greater calorie output would also occur by exercising a little harder or longer (around ten minutes would do), and thereby avoiding the unpleasant sensations associated with sweating in heavy clothes. During my search, I inadvertently stumbled upon a link to an article describing *7 foods that'll naturally boost your body temperature*, suggesting that it isn't just clothing manufacturers exploring this line of enquiry. It's a tautology that consuming calories to burn calories is somewhat counterproductive. Even to assume a charitable stance and accept the 100-calorie claim verbatim, there are a plethora of other ways one could manipulate their calorie expenditure by this amount. For my second attempt, I contrived to think more laterally.

2.4.2. The O_2 Drink

This product is a liquid drink one can buy at a convenience store. Instead of adding carbon dioxide (CO_2) to the mixture making it carbonated, O_2 is added which invigorates the liquid it with oxygen. When consumed, oxygen diffuses through the small intestine and into the blood where it binds to hemoglobin – the oxygen-carrying component – which, in turn, increases oxygen delivery to exercising muscles and the brain. The O_2 drink could be taken at any time to provide a dose of vitality and wellness, but it'd be particularly potent during periods of the day typically associated with lethargy and low energy, like after lunch, thereby improving concentration. There'd be a special formulation consumed immediately before or during exercise as an ergogenic aid. Young professionals and city workers would be prime consumers, struggling to meet their deadlines, burning the candle at both ends, and trying to balance a hectic work schedule with an equally hectic social life.

The closest product available is one with the trendy street name *Oxigen*, which is purportedly enhanced with *powerful O_4 oxygen molecules*... getting you *ready for faster recovery and more stamina.* My search also returned information regarding oxygen gas shots which are widely available in some countries at equally bourgeois oxygen bars, but these are more associated with hedonistic experience rather than fitness or health. Searching for *oxygen drink* inadvertently opened a rather large can of worms. The *oxygen cocktail* is a foamy drink enriched with gaseous O_2 which is used as an adjunct to oxygen therapy by Russian medical institutions. In the latter, doctors insert probes into the patient's stomach, through which the body could be filled with up to 2 liters of oxygen. Although patient outcomes were reported to improve with the therapy, the procedure was abandoned due to discomfort caused by the probes; later, it was decided to formulate a drink. It's claimed that oxygen cocktails have numerous positive effects including reducing fatigue, improving sleep, activating metabolism (although I'm unclear of the meaning of this phrase), and boosting the immune system, although the contraindications are equally extensive. The extent of my research reveals it as a distinctly Russian therapy, with very little published data, and seemingly none outside of Russia.

Not unrelated to oxygen drinks, energy-boosting green chlorophyll smoothies are a thing. Chlorophyll is the green pigment of plants, and its primary function is absorbing light for photosynthesis, the chemical reaction in which plants combine CO_2, water, blue and red light as ingredients for the synthesis of sugar which is used as food. The premise is that ingesting chlorophyll can somehow improve oxygen retention and/or delivery in the human body but, according to our biological understanding, it can't. Nevertheless, one creates a smoothie by blending predominantly green fruits and vegetables (kale, celery, spinach), and liquid chlorophyll can be added to the mixture directly to enhance its concentration of green chlorin pigments. One website claimed that chlorophyll *helps in treating anemia, may fight cancer cells,* and *contains high levels of vitamins and minerals*, claims that go completely unfounded in the scientific literature.

It's worth noting that healthy individuals (i.e., non-patients) are perfectly well developed to diffuse oxygen through the lungs and into the blood where it gets transported to the muscles and organs of the body. Unless environmental stimuli (like high altitude) preclude it, oxygen saturations will generally remain close to 100% at rest. It was long thought that yawning served the purpose of transiently augmenting oxygen delivery to the brain – an unfounded fabrication that propagated due to hearsay – but yawning is now known to serve no such purpose, and most likely has value in social cohesion. Taking oxygen by almost any other means is unlikely to provide any additional benefits. Even providing athletes with a brief, transient burst of supplemental oxygen gas (via a face mask) fails to induce an ergogenic benefit because: (1) oxygen diffusion through the lung and saturation into the blood are rarely limiting factors to exercise performance in healthy people; rather, performance is limited by our capacity for oxygen *delivery*; and (2) any modest effect of supplemental O_2 will dissipate moments after the gas mixture is removed. Accordingly, we can discount these unscientific myths, along with any product that propagates said myths in a cynical attempt to monetize oxygen.

2.4.3. The Intravenous Protein Pump

At the last, I tried to design a product that exceeded my earlier boundaries of scientific implausibility. It's a bag of water or saline, mixed with essential amino acids and nutrients, and then drip-fed via an intravenous cannula in the antecubital vein into the blood stream. Because the cannula bypasses the gut, it delivers nutrients faster than if taken orally as food or tablets, and there are few calories consumed. The pump would be marked with a slogan akin to *why suffer the inane inconvenience of eating and drinking, when you can infuse your nutrients directly*; or perhaps, *give your gut the vacation it needs.* The target audience is again rather broad; this isn't a medical device for patients critically deficient in nutrients, nor is it specifically a treatment for anemia in which a doctor may prescribe intravenous iron administration. It's for healthy people, specifically athletes, exercisers, and young professionals too busy to concern themselves with healthy-eating or regular exercise. The real market would be large businesses which would fund their employees to undergo the therapy on a bi-weekly basis as part of a periodic detox program. Indeed, characterized by long working hours and high-stress jobs, the corporate world would be seen prioritizing the health and wellbeing of its workers with this trendy new infusion regimen, thereby improving their productivity.

I was surprised to learn that such a product hadn't escaped manufacture; with several locations in London, UK, and similar outlets in the US, *GetADrip* offers intravenous drips and booster shots to the general public to improve a number of first-world complaints. The business exploits our hardwired yearning for *quick fix fitness*, but lacks any published literature to support its effectiveness. Their website comprises a *menu* of treatments, complete with bolt-ons and upgrades, each more fantastic than the last. There's a *Party Drip*, an *Energy Drip*, an *Immunity Drip*, a *Detox Drip*, a *CBD Drip (cannabidiol)*, a *Beauty Drip*, a *Fitness Drip*, an *Anti-Ageing Drip*, a *Slim Drip*, and the list continues. A button-click opens an ingredients list for each treatment. At the time of writing, the Fitness Drip contained: L-carnitine, B-vitamins, saline, potassium, calcium, bicarbonate, and glutamine. The ingredients of the Beauty Drip were: B-vitamins, bicarbonate, calcium, potassium, saline, glucose, glutathione, and vitamin C. The Detox drip had: glutathione, vitamin C, B-vitamins, saline, potassium, calcium, and bicarbonate. The Slim Drip comprised: potassium, B-vitamins, amino acids, methionine [also an amino acid], L-carnitine, calcium, and bicarbonate. There's a conspicuous trend that these intravenous treatments contain a standard list of six or seven ingredients, the order shuffled, and with a few minor additions or omissions that purportedly confer the principal characteristic of the drip. Amino acids supposedly make you fit, while L-carnitine makes you slim, neither of which has any empirical basis. Precisely what distinguishes the *Beauty Drip* from the *Fitness Drip,* aside from L-carnitine and 1,500 mg of L-Glutamine (you'd obtain three to four times this amount of L-Glutamine in a chicken breast), is unclear. I discussed these distinctions with a representative of the brand, but he was unable to provide any insight beyond *L-carnitine will activate your metabolism*, a claim without a cogent meaning. Yet, it's all inconsequential amidst a trendy New-Age façade, helpful assistants, and a menu of colorful picture icons denoting the characteristics of each drip. Shrewdly, the company carefully *implies* benefits without overtly stating them. In fact, nowhere on the site is it stated that *this drip will improve your fitness*, or *this injection will make you stronger.* The moment a lucid statement for efficacy is made, the claim becomes testable, and a predictable conclusion will result. But the site is designed in such a way as to encourage the consumer to *infer* a benefit.

Notwithstanding the lack of evidence that the ingredients of a given drip confer any of the implied benefits, from a scientific perspective, contrivances like this are superfluous for several reasons. First, your gastrointestinal tract has evolved very efficient digestion and absorption mechanisms that allow nutrients to pass from the gut and into the blood. There are exceptions in which a medical condition might preclude this and nutrient infusion may become necessary, but these products aren't aimed at patients and besides, such therapies would be administered by a physician under controlled conditions. Moreover, we later review evidence showing no advantage of intravenous infusion over oral ingestion of nutrients in healthy people. Second, multiple lines of research have shown that vitamin and/or mineral supplementation confers no benefit if one follows something resembling a healthy diet. Only in very particular cases, e.g., if you were diagnosed with a nutrient deficiency that couldn't be resolved through dietary means, would supplementation be necessary.

I thought myself worldly, with experience aplenty in these domains of science and *pseudo*science, but in my research for this book and with exercises such as this, my own thresholds have been heavily taxed. As I concocted scientific mechanisms, and retrofitted them to fabricate a bogus sales pitch, I was struck at how easy it was and also how much fun. The factual basis, accuracies or inaccuracies of my statements were of little consequence, and it was pleasant being liberated from the responsibility of having to conceive a product that might actually be effective for its intended purpose. I was concerned only with inventing an idea that would be marketable. I concede that the examples presented here are extreme in nature but, nevertheless, they're for sale online and in the stores. In the coming chapters, we tackle far more complex and deserving topics that make larger claims and on a broader scale, and we delve more deeply into the logic and evidence that underpin them; indeed, not all products are as easy to dismiss as those aforementioned.

As we progress, it will become intuitive to distinguish between the fake and real treatments. The former generally rely on three components to exert a positive effect: (1) something resembling an authority to lend legitimacy to the treatment; (2) an intervention of some kind, perhaps a pill or an injection; and (3) the consumer's pre-existing belief that the intervention will exert an effect. But a *real* treatment, something that confers a genuine physical outcome, will work regardless, and of its own accord. An infant sick with a bacterial infection has no concept of a medical authority, and no pre-existing expectation, but antibiotics will facilitate their recovery; a broken bone, if set in place and immobilized, will repair and re-strengthen in time; an exercise regimen that involves frequent aerobic activity will provoke cardiovascular adaptations that improve aerobic power, leading to better health and endurance performance. These are all objective responses, testable and quantifiable, independent of placebo or bias or belief.

3

LOGICAL FALLACIES IN SPORTS SCIENCE

fal·la·cy.

Incorrectness of reasoning or belief; erroneousness. The quality of being deceptive.

3.1. Playing by the Rules

Take a few moments to recall your school days. This might be easier for some than others; indeed, some of my undergraduate students left school only a few years ago. Think about the actions in which you engaged during the daily lunch break, and whether you chose to pursue any after-school extracurricular activities. There's a good likelihood you competed in sport (e.g., track, football, netball, basketball, hockey), or perhaps it was something more academic (e.g., chess or debate club). Successfully engaging with other students from your school, or others in the state, was a relatively easy process; once the basic logistics of time and place had been determined, it was as uncomplicated as turning up and starting play. There was no need for convoluted discussions or organization, and all held a common understanding of the overarching rules and regulations, as well as the finer nuances. This shared understanding is a prerequisite to a fair and mutually enjoyable game, and assuming everyone abides by the rules, there will generally be a conclusion (win, lose, or draw) on which everyone will agree (or at least, that's the purpose). Consider how frustrating it would be if an opponent sat opposite you behind the black chess pieces, only to insist on moving first. Think how unjust and one-sided your soccer game would become if you were passing and dribbling with the ball, shooting at the goal, while your opponents were scooping up the ball in their hands and rapidly advancing down the field. What about if the runner in lane 6 commenced his sprint half-a-second before the starter pistol, without repercussions. If your opponents didn't understand the rules, or perhaps weren't concerned with abiding by them, your team might be crushed 27–0 without hope of reprieve or reprisal. It may be an empty victory – certainly subjective rather than objective – but it'd be a victory nonetheless.

The same can occur in logical discourse. If, in a rational discussion aimed at establishing *truth*, you are playing by a different set of rules to your opponent, there's a strong likelihood that logical discourse will break down. Recall that the sale of a product is usually accompanied by a claim, or series of claims, regarding the product's effectiveness; these claims are confident and forceful statements of fact or belief. Some examples are: *made with all-natural ingredients; over a million units sold; recover faster; a traditional technique practiced for hundreds of years; scientifically-formulated.* The factual basis of the claim must be scrutinized because we need to establish if it's likely to deliver on the promise, and this is our focus in the next chapter. Presently, however, it's the anecdotal testimonials, misleading data designed to bias our opinions, and the vague, unfalsifiable statements pertaining to the product that are under inspection. The supporting statement – the sales pitch, the rhetoric – also deserves scrutiny to ensure it follows sound reasoning and consistent logic since, when logical discourse is fractured, the assertions are usually rendered invalid.

For practitioners, exercisers, athletes, and coaches, assimilating the detail of this chapter is critical in understanding the differences between valid and invalid reasons for buying a product. This will be an important string to your bow, a tool in your toolbox, enabling you to better navigate the complex world of commercial health and fitness.

3.2. The Logical Fallacy

A logical fallacy is an error (accidental or deliberate) in reasoning. Fallacies are tricks or illusions of common sense; they can be smuggled into an argument to deceive you into taking favor with a particular perspective, or you might inadvertently commit a fallacy in your own interpretation of some information. The fallacies themselves appeal to our basic experiences of being *sold* something. As previously discussed, the notion of a *sale* doesn't only denote a monetary exchange, but any instance in which someone might try and persuade you to a cause in sport, be it a training program, a supplement, an opinion, or an ideology. For this reason, logical fallacies appear frequently in debates. In his seminal book *The Demon Haunted World*, cosmologist and legendary science communicator Carl Sagan described why comprehension of the logical fallacy is important:

> ...any good baloney-detection kit must also teach us what not to do. It helps us recognize the most common and perilous fallacies of logic

and rhetoric. Many good examples can be found in religion and politics, because their practitioners are so often obliged to justify two contradictory propositions.

This section will further hone your skeptical outlook and increase your sensitivity to poor arguments, as are frequently proposed in the sale of sporting goods. There exist a great many logical fallacies, each with their own subtle details and nuances. In the health and fitness industry, a small number rear their ugly heads time and time again to distract and deceive. I've condensed the list to the nine that I consider the most frequently committed, and have offered clear sporting examples for each. As we progress, recall the products you've seen or tried, and whether they were sold to you on the basis of such a false premise.

3.2.1. The Appeal to Popularity

Aka. *argumentum ad populum*; appeal to numbers; appeal to the masses; (jumping on) the bandwagon fallacy. The appeal to popularity is perhaps one of the oldest appeals to serve the commercial machine; the fallacy is to assume an assertion true because many, or a majority, of people agree with it. It's considered misleading because an assertion can be erroneous regardless of how many people are in favor of it. Consider an activity tracker that's marketed with the slogan *over 1-million units sold, worldwide* or perhaps a supplement that's branded *Britain's most popular protein*. It stands to reason that the number of people purchasing the product says nothing of its quality or effectiveness, since popularity can be influenced by clever marketing, creative statistics, political control, or a falsification of data. An appeal to popularity with which we're all familiar is the online product review; we're more inclined to buy something if it's been positively rated by a substantial number of our peers because it appeals to our numbers bias. But humans have the propensity to convince themselves of efficacy where there commonly is none, via the placebo effect, and various forms of confirmation bias. Accordingly, a product must be judged on merit alone – a tautology to which we'll be returning frequently – and popularity shouldn't replace independent, third-party research.

3.2.2. Blinding with Science

The aim of committing this fallacy is to overwhelm someone with technical details and jargon in order to deliberately bemuse or mislead them. Most commonly, this fallacy invokes science-sounding terms with which the consumer is unfamiliar, thereby instilling the product with a false legitimacy. On occasion, the product description is so complex, overflowing with elaborate terms, you'd need the *pseudo*science incarnation of the Rosetta Stone to decipher its meaning. Few products are sold alongside as much outright nonsense, flawed logic, and false claims as the PowerBalance bracelet. Briefly, PowerBalance was marketed with the slogan *holographic technology embedded with frequencies that react positively with your body's energy field.* There isn't a chance of deriving any meaning from this statement which is largely *pseudo*scientific. The term *holographic technology* is nonspecific, and the notion that a hologram could *resonate with frequencies of the body* is not a physical phenomenon. The terms engender something vaguely *sciency*, but have little meaning or application in the real world.

An altogether subtler means of blinding with science is to employ technical terms that relate to real mechanisms, but either misrepresent those mechanisms or, worse, get them dead-wrong. This might be to deliberately deceive, or it might be inadvertent and the consequence of poor understanding. An ongoing frustration in the exercise science community is the misappropriation of lactic acid. No two words are capable of striking more fear into the hearts of athletes and coaches than lactic acid, owing principally to a pre-1980s understanding of human physiology, a time when students were instructed that lactic acid accumulated in the muscles during high-intensity exercise, causing pain and fatigue. The term is now synonymous with fatigue and has become entrenched in everyday sporting language. Athletics clubs operate *lactic* sessions in which track athletes run as fast as possible to accumulate lactic acid and become robust to its ill-effects. I recently ran a poll with 225 undergraduate sports science/kinesiology students, with an overwhelming 90% of them of the misunderstanding that lactic acid caused muscle fatigue during exercise. And it's no surprise, because it's ubiquitous with contemporary sports culture. During the 2018 Rio Olympic Games, Sir Chris Hoy – sports ambassador of Great Britain, hero of Scotland, multi-time Olympic Gold medalist – cited lactic acid and its fatigue-inducing effects on at least three separate occasions during the track cycling competition. It's surprising to hear this from someone who resided in a high-performance program for over a decade. In any case, lactic acid, more specifically its derivative *lactate*, is an important energy source. The work of George Brooks identified that during sustained exercise, approximately 75% of lactate is oxidized immediately for energy during muscle contraction, while the remaining ~25% is shuttled to the liver and converted back to glucose where it can make a continued contribution to energy metabolism[8]. Lactate is critical to prevent pyruvate accumulation which, in turn, is critical to supply the metabolic component for continued breakdown of glucose (termed glycolysis). If the muscles didn't produce lactate, acidosis and muscle fatigue would occur *more quickly*, and exercise performance would be impaired[9]. These aren't sensational pronouncements for anyone in my industry; physiologists have had a robust understanding of these phenomena for nearly four decades. Despite the availability of these data in the public domain, it doesn't stop product manufacturers from capitalizing on lactate's bad reputation to sell product – compression garments, tights and socks, contrast water therapies, ice bathing, altitude interventions, and supplements – all purported to attenuate lactate accumulation and/

or facilitate its removal during or following exercise specifically *to reduce muscle pain and fatigue*. It's easy to demonize the devil, and use the resulting fear to further your cause. This is a prime, albeit nuanced, example of the *blinding with science* fallacy, since most in the sports industry know enough to conflate lactate with fatigue (thus giving credence to strategies that claim to mitigate the pair), but not enough to see through the factual error.

Some adverts cite more general phenomena like *subatomic energy*, and reference Einstein's theory of Special Relativity. Terms like *quantum*, which have meaning only in particle physics, have been hijacked by *pseudo*scientists selling products on the basis of the *blinding with science* fallacy. Consider that products in many facets of society (e.g., sport, cosmetics) are labeled as *scientifically formulated* to imply an endorsement by science. But be on alert; the overuse of technical jargon, or vague and nondescript scientific terminology, could mean that you are at risk from this fallacy.

3.2.3. The Appeal to Antiquity

Aka. *argumentum ad antiquitatem*; appeal to common practice; proof from tradition; the appeal to tradition. Used with increasing regularity in the sports industry, the appeal states that an assertion is correct because it correlates with a past practice or tradition. It implies that the traditional practice might be valid because it hails from a *golden prior age* when athletes were naturally trained and sport was pure, and similarly that contemporary strategies are complex, ineffective, and contaminated by modern science and technology. Traditions are passed down from generation to generation, remaining intact often for no other reason than engrained behaviors or sentimentality. The fallacy is contrary to the obvious notion that, in exercise and sport, the only appropriate reason to implement a new practice, or maintain an old one, is because it's *effective*. Many institutional sports with a strong historical influence and harboring a rich tradition – e.g., soccer, boxing, or martial arts – often maintain antiquated and sometimes discredited practices because coaches are passing onto their athletes the techniques taught to them by *their* predecessors, without independently verifying the effectiveness of the practice. It becomes a self-perpetuating sequence of misinformation. Clearly, breaking the shackles of tradition is no easy feat.

Two prominent examples of the *argumentum ad antiquitatem* in sport and exercise are the re-emergence of barefoot running, and the growing popularity of the Paleo Diet. Both these systems stake a claim for efficacy, at least in part, because they correlate with a past tradition, making the assertion that the practice is valid on this premise. Barefoot running is covered more comprehensively in Chapter 8, so for now I'll scrutinize the Paleo Diet. The term *Paleo* is derived from *Paleolithic* which denotes the early phase of the Stone Age. This commercial diet, originally dubbed *the caveman diet*, promotes foods that were allegedly consumed by our ancient Paleolithic ancestors – meat, nuts and seeds, vegetables and berries – while omitting those that are the consequence of modern agriculture, such as dairy, refined sugars, and starches. The recommendations are generally sound, although the hearty consumption of red meat that characterized the early incarnations of Paleo has been linked to an increase in the risk of cardiovascular disease on a population basis[10]. There are also definitional inconsistencies with Paleo because a *traditional* Stone Age diet would have comprised insects, roots, flowers, stems, and the meat, cartilage, organs, and bones of the animal, facets of the diet that are conveniently discarded to make it more commercially viable. Nevertheless, if the diet is to be followed, then it must be on the basis of sound nutritional recommendations (i.e., on merit), and not on its fallacious claim to traditional practice which holds no sway in a scientific argument.

Another, perhaps less obvious, example of the appeal to tradition is in the claim that distance runners must engage in very high-mileage training in order to maximize their chances of success. To lend legitimacy to the contention, the practices of very successful Kenyan and/or Ugandan distance runners are referenced, further bolstered because most African distance runners begin running at a very young age, often because of culture or necessity, largely forgoing the scientific approach due to a lack of facilities and/or available expertise. It's likely that this type of training is, in fact, highly efficacious for distance runners, and there is research suggesting that high-mileage training – by manipulating glycogen content – might promote endurance adaptations via the stimulation of various cell-signaling pathways[11]. There's even the suggestion that African distance runners perpetually train in a slightly nutritionally-depleted state, which encourages fat utilization and endurance adaptation. Regardless of the scientific efficacy, to assume that such training is beneficial, purely because it correlates with the *traditional* practices of a successful populace of native African runners, is to commit the fallacy of an appeal to antiquity. As a caveat, it would also be unwise to assume that a practice is inherently wrong because it correlates with a tradition, or because its premise is based on a fallacious appeal. Again, the salient point is that the practice be valued on evidence and merit, without harboring favor or disfavor because of its historical use.

3.2.4. The Argument from Authority

Aka. *argumentum ad verecundiam*; appeal to authority. Another frequently committed fallacy, the argument proposes that a position is true, or more likely to be true, because an authority supports or endorses it. High-profile athletes are often recruited into advertising campaigns to espouse the wonders of any number of health and fitness products. It's anticipated that respect and reverence for the athlete might confer a similar respect and reverence for the product, thereby facilitating sales. Nutrition or sports garment companies often sponsor successful athletes. For example, Lance Armstrong was sponsored by Nike as far back as 2004; he wore their clothes and they produced the yellow wristbands of the Livestrong brand. Yet, despite the close affiliation, it would be unreasonable to conclude that Nike products contributed to Armstrong's status as the world's most revered cyclist. But by affiliating themselves with Armstrong, Nike affiliated themselves with success. And when insurmountable evidence emerged that Armstrong had competed using banned substances, his sponsors dropped him like a bad habit.

Soccer boots are advertised by some of the world's greatest and most revered players; Adidas Predators have been endorsed by David Beckham, Xavi, Robin van Persie, Michael Ballack, Raúl González, Steven Gerrard, Dirk Kuyt, Kolo Touré, Edwin Van Der Sar, Petr Čech, Brad Friedel, Anderson, and Zinedine Zidane. While these high-profile athletes may have an affinity for the product, it'd be a stretch to attribute their extraordinary skills to the leather and rubber adorning their feet. Michael Jordan sells Gatorade, Mo Farah sells Quorn, and Jess Ennis sells credit cards, skincare products, and PowerAde. There even exist websites that are designed exclusively to assist marketing departments select the most appropriate athletes to endorse their brand. As critical-thinkers, we simply cannot accept these endorsements at face-value. When the appeal to authority is invoked, there are some pertinent questions to be posited: *what does the athlete really know about this product? Is the athlete sufficiently qualified to be endorsing this product, and what are the athlete's credentials? Have they tried every soccer boot on the market and made an objective, informed decision? Have they sampled every carbohydrate supplement? Would it matter if they had done so, given that every athlete has a different opinion on taste, style, and fit? How much do you think they got paid to wear that?* I'm not sufficiently cynical to suggest that athletes endorse products only for monetary remuneration, and either way it's inconsequential. If an athlete exploits their status to facilitate the sale of a product about which they're not a legitimate expert, this can be considered a fallacious argument from authority.

An important nuance of this fallacy is when a position is endorsed by a legitimate authority in a given field (e.g., a doctor or scientist) but whose credentials are not appropriate to the case. If an individual is not an authority in the topic on which they're commenting, or they're not an expert at all, then their opinion isn't valid. Be mindful that even a world-renowned expert on a discipline can make a mistake or affirm a falsehood; nobody is infallible, and no testimony of *any* authority is guaranteed to be true. It's for this reason that I encourage my own students to independently question and verify for themselves the statements of the so-called experts, including their tutors, including me. Carl Sagan once wrote:

> One of the great commandments of science is, 'Mistrust arguments from authority.'...Too many such arguments have proved too painfully wrong. Authorities must prove their contentions like everybody else.

That is the beauty of science. Under these circumstances, we must consider the authority and decide upon their appropriateness for the situation. More importantly, we must evaluate the statements made, and decide on the correctness of those statements by researching them objectively.

Consider this from another perspective; if you had a sports injury sustained in competition or in the gym, would you seek advice from someone with a PhD in geology? Would you consider your friend who's a biology expert to be more appropriate? You could visit your physician, and for many people, this is the first port-of-call. Ultimately, however, you'll get the best advice from a sports injury specialist. It'd be wrong of me to call upon a false authority to help with my swollen ankle just because they had a keen reputation as a scientist or generic intelligence. A world-renowned expert in one domain is precisely that – an expert in one domain – and I'd be no more willing to accept medical advice from Albert Einstein than I would from a patron of the local pub.

Finally, I urge caution on absent-mindedly challenging an authority. Our culture of post-truth and pervasive social media has elevated all platform users to expert status. Experts in any subject must be challenged, but preferably by others who have diligently studied the field, the literature, and who have developed over many years a theoretical and conceptual understanding of the topic. Don't plunge into a debate for which you're not appropriately credentialed; not all opinions are equally valid.

3.2.5. The Appeal to Nature

This is perhaps the most widely committed fallacy in all of sport, health, and exercise. In the appeal to nature, one proposes that something is *good* because it's natural, and *bad* because it's unnatural (or synthetic). It underpins every product advertised as being made with *all-natural ingredients*. The proposition is invalid because it's founded on subjective opinion as opposed to objective fact. Furthermore, the stance is intellectually meager because there's no lucid definition for *good* or *bad* in this context, and no coherent argument for favoring something natural over something unnatural. The entire organic food industry is founded on this fallacy. It's relevant for the health and fitness industry because organically grown produce makes an overt claim for being more healthful than that non-organically grown; this would benefit a healthy lifestyle or an athlete-in-training if the claims were true. On the contrary, the claim has been studied extensively, and is unfounded. The appeal to nature is not so much an ethical issue as it is a semantic one.

I recently noticed a high-street store which had the words *Natural Health* emblazoned above the door in sizable green writing. Through the window, I saw various wholefoods, supplements, and diagrams depicting the musculoskeletal system, energy crystals, and burning incense. Take a moment to consider what's implied by the name *Natural Health*, whether it's an alternative to *Unnatural*

Health, and why *natural* health should be preferred to *unnatural* health. Far from being inane questions, they're essential for challenging the way we conceptualize health and fitness. Assume that the outcome of shopping at this store is improved health or wellbeing. Now, imagine there are two stores side-by-side on the promenade: one of them called *Natural Health* with a big green sign and a logo of a tree, and the other called *Unnatural Health* with a blue sign and a logo of some tablets. The outcome of making a purchase at either store would be equivalent, i.e., an absolute improvement in *health*. Given this assumption alone, which store would you choose? If the former, ask yourself why. All things being equal, there's no coherent reason to favor one store over the other; yet, a majority of people would opt for the natural alternative any day of the week.

A recent discussion with a colleague from the university, whose expertise lies in exercise referral and cardiovascular health, framed the fallacy in a novel light. We encountered the subject of pharmaceuticals that are prescribed by doctors for one or more lifestyle-related ailments. We concurred on the notion that *exercise is medicine*; if there were a drug available that could bestow the individual with all of the same benefits achieved by regular exercise, its maker would fast be the recipient of a Nobel Prize. The long-term benefits of regular exercise and physical activity are well documented, and so are the benefits of statins and ACE inhibitors, prescribed to lower cholesterol and blood pressure, respectively. I argued that, assuming the same outcome, it shouldn't matter whether one lowers their blood pressure via regular exercise or by taking ACE inhibitors. Now in reality, there are a number of reasons why exercising to achieve this outcome is preferable; for example, exercise is also important for maintaining good cardiovascular health, improving insulin sensitivity, functional strength, meeting friends, social cohesion, etc. Drugs are also costly, and often associated with side-effects. Nevertheless, there are some individuals who can't exercise because they're too unwell, they have chronic arthritis, or a spinal cord injury, or their blood pressure is dangerously high. In these instances, drugs may be preferable. There's simply no logical or theoretical reason to prefer exercise over drugs to lower blood pressure alone on the basis that exercise is *natural*, especially if arriving at an equivalent outcome.

To cement the point, take a glucose molecule synthesized in a lab. It has the chemical formula $C_6H_{12}O_6$, which means that it contains six carbon atoms, 12 hydrogen atoms, and six oxygen atoms. A molecule of naturally occurring glucose (created in nature) also has a chemical structure $C_6H_{12}O_6$, and if you compared the two molecules under an electron microscope, they'd be indistinguishable. So, the pertinent question is: should the natural sugar be given preferential treatment over the synthetic sugar, and is it considered more healthful or ethical to consume? The logical answer is no; they're chemically identical and will identically influence physiological outcomes when consumed. To think otherwise is an example of the *appeal to nature*; yet, our inherent bias for natural products is pervasive in the health and fitness industry, another exploitable flaw in cognition.

3.2.6. Definitional Ambiguity

Aka. persuasive language. This type of fallacy takes the form of misleading or vague language to deliberately deceive, or to make a claim easier to defend, and it's a relative staple. In logical discourse, the words chosen to convey a perspective should have a conventional meaning on which all can agree, but in product advertising, there's often tremendous ambiguity in those commonly used. For example, the word *fast* is applied in the context of human movement, and has a concise definition in the Oxford English Dictionary: *moving, or capable of moving, at high speed*. By extension, the term *faster* would imply a relative improvement in the rate at which something moved or was capable of moving. So, if a running coach assured you that their new training program would make you run a *faster* marathon, you could choose to accept or reject their claim on the basis of the generally accepted definition. In the sporting context, other expressions are less explicit. The term *recovery* is used frequently in the marketing of supplements, recovery shakes, massage therapy, stretching, cryotherapy, ice bathing, etc., but has no definitive operational definition or end-point. Does it denote a return to *baseline* function? Does a full recovery include the supercompensation that results from successive training sessions? Does recovery include all facets of biological function, or is it specific to the musculoskeletal system? What about recovery of immune properties, hydration, or psychological wellbeing? By itself, the term *recovery* is ambiguous in that it cannot pertain to all bodily functions and, given the frequency with which the term is used, perhaps requires clarification in each context. Other terms like *wellness* are ubiquitous with the health and fitness industry, and yet *wellness* has no cogent operational meaning. Indeed, it can be interpreted to serve any number of functions from person to person, and that's the intention.

When ambiguous wording is invoked in a claim, the door is left wide-open for a definitional retreat when the claim is challenged. The meaning of the word can be amended to deal with an objection raised against the original assertion. For example, a product that specifically claims to make you *run faster over 400 meters* can easily be tested; there's little ambiguity because the language is detailed and specific. By contrast, a product that claims to *improve wellness* is far easier to defend when challenged, whether in a friendly debate or in court.

Definitional ambiguity can also be used to great effect to imply a claim without explicitly stating it. Let's say that after increasing the intensity and/or frequency of your training, careful analysis of your dietary requirements reveals that it's time to invest in a whey protein supplement to help you satisfy your protein needs. There are literally hundreds of brands available, each claiming precedence over the other. Next, you see a full A4-page advert in a glossy fitness magazine for *Tiller's Whey Protein – Britain's Leading Protein Supplement*. If you choose to accept the assertion made in the advert – that my protein supplement is Britain's leading brand – you're reasoning as follows:

1. Tiller's Protein is the leading protein supplement in Britain.
2. If you purchase a whey protein supplement, then you should choose Britain's leading brand.

These are valid assumptions if we're playing by the rules of logical discourse. However, the advert commits the fallacy of definitional ambiguity. What exactly does *leading* mean in the context of a whey protein supplement? Indeed, it's sufficiently malleable. It could mean that it's Britain's most *effective* whey protein in that it most effectively stimulates muscle protein synthesis; and if this were the case, you'd probably demand some evidence to prove the assertion. However, the term *leading* could also denote that Tiller's brand is Britain's *largest* whey protein supplier, or perhaps it's the best-selling, or even the highest grossing (most profitable). This distinction is paramount because, after all, the fact that a protein supplement makes more money than another is hardly testament to its effectiveness, and scarcely a reason to invest. I'd sooner buy an unbranded supplement I knew to be potently effective, than a popular product that was less so. Choose carefully your words and scrutinize closely those employed by others; if you infer a meaning that wasn't implied, then you're at fault, and the consequences yours alone to bear.

3.2.7. Confusing Correlation with Causation

Aka. *post hoc ergo propter hoc*; *post hoc* fallacy. This fallacy of interpretation and pattern-seeking can strongly bias our opinions, and it's commonly committed by the consumer of a sports product. In his book *Why People Believe Weird Things*, Michael Shermer postulates that humans have developed a general belief engine:

> We evolved to be skilled, pattern-seeking, causal-finding creatures. Those who were best at finding patterns – for example standing upwind of game animals is bad for the hunt, cow manure is good for the crops – left behind the most offspring. We are their descendants. The problem in seeking and finding patterns is knowing which ones are meaningful and which ones are not; unfortunately, our brains are not always good at determining the difference.

The *post hoc* fallacy is committed when one observes two events occurring in sequence, one after the other, and then asserts that the first event must have caused the second; it's a form of superstition. The mistake is to assume correlation between the events when there could be several ill-considered explanations for the second event. Here are some examples: a basketball player dons a particular brand of underwear, happens to perform phenomenally well in an important game, and assumes his performance was attributable to the garment, and now wears his lucky briefs for every match; a soccer player scores a hat-trick, a performance she attributes to her new football boots worn on their first outing; you buy your netball team a new water dispenser at the gym, you see their performances improve for the remainder of the season, and reason *post hoc* (after the fact) that the players must have been dehydrated all along. Consider some lower-limb compression garments that are often worn following a tough training session in order to promote recovery. After the session, you squeeze your tired, dilapidated limbs into some tights and wear them to bed. The following day, you rise feeling pain-free and refreshed, and assume that your expensive new tights were responsible. Your compression tights may well have expedited your recovery but not necessarily so; to assume that link – that the one action necessarily resulted in the other – without objective evidence-for-efficacy – would be to commit the *post hoc* fallacy. After all, there are many other factors that could function to facilitate your recovery, such as the food you ate that evening, securing a long and restful night's sleep, or perhaps you took some powerful supplements or anti-inflammatory drugs. It's plausible you'd have woken feeling refreshed and pain-free in any event, regardless of the intervention. Failing to consider the alternative hypotheses is a type of heuristic called a congruence bias.

It's quite common for exercisers to commit this fallacy because humans prefer to have explanations for the outcome of events, even if it means observing an event and then retrospectively attributing a cause. Human history is littered with instances in which we've contrived answers to previously unexplained events, rather than having the humility to live with the unknown. The fallibility of the correlative assumption is keenly exemplified online at www.tylervigen.com. The site's author takes two unrelated variables and plots them graphically so that they appear inexplicably linked; he then performs some basic statistics to examine the strength of the relationship. For example, the number of people who drowned by falling into a swimming pool correlates strongly with the number of films in which Nicholas Cage has appeared. An r value of 1.00 would signify a perfect correlation between the two variables, and those in the example correlate with $r = 0.67$. It's a crude but effective illustration of how the *post hoc* fallacy can result in spurious decisions, since canceling Nick Cage movies is unlikely to reduce the number of deaths related to pool drowning.

A research colleague of mine once admitted falling for the *post hoc* fallacy. He disregarded all complementary medicine with the exception of acupressure (the practice of applying pressure to various body parts in order to facilitate energy flow and reduce pain). His flawed reasoning arose because his partner's back pain once improved overnight following application of acupressure to the balls of her feet. Such practices are widely discredited for being unscientific, and a more likely explanation is that the issue spontaneously resolved, as is the nature of acute back pain. At least spontaneous recovery is scientifically plausible. Given that humans have been deceiving themselves in this manner for many millennia, you'd be forgiven for exhibiting some semblance of frustration; how can we know the true cause of events if our inherent faculties are so untrustworthy? Certainly not by intuition, or sole testimony, or anecdote alone, but rather via objective investigation of the phenomena, preferably in studies designed to control for such confounding variables.

3.2.8. The False Dichotomy

Aka. the all-or-nothing fallacy. When you're presented with only two options, but more options clearly exist, then there is an attempt to obligate your decision via the false dichotomy fallacy. With this informal fallacy, you're presented with an either/or option (with only two outcomes); it may be that you've inadvertently omitted a third, or weren't granted it in the first instance. When this fallacy is committed intentionally, it's because someone is trying to force from you a choice; e.g., perhaps one of those choices is deliberately unappealing. Very effective in advertising, media, and politics, advertisers often present black-and-white thinking when they subtly suggest that you follow *this* program, buy *this* product, follow *this* diet, or be fat and unhealthy, as if there weren't any alternatives or spectrum of possibilities between the two extremes. In this industry, there's sometimes a subtle undertone that one must aspire to be a professional athlete or completely sedentary, when there's actually a broad spectrum of physical fitness along which you'll reside. Attaining professional athlete status is an unreachable ideal for the vast majority.

The false dichotomy can also manifest more subtly in the scientific literature. Studies may observe a course of yoga to evoke effective long-term weight management. This doesn't mean it's necessarily *more effective* than other exercise interventions (you could lose weight playing soccer, running, lifting weights, etc.) but these alternatives are rarely presented. Researchers are obligated to espouse the novelty of their findings; it confers a greater chance of being published, and likely increases the number of citations or media coverage received. Concluding that *yoga was shown to facilitate long-term weight management, but is likely no more beneficial than any other form of exercise* might be a reasonable conclusion for an objective review article, but for an original manuscript, it doesn't exclaim *unique finding*. It could be considered a false dichotomy to present the data as if yoga were the only means of losing weight.

One area of study in which I'm personally active is the pathophysiology of ultra-marathon. There's sufficient data now to suggest that long-term participation (over many years) might lead to negative health implications, particularly from a cardiovascular perspective, and particularly in susceptible people. I've received a little criticism from those who believe that the risks of poor health from long-term ultra-marathon running are far lower than those posed by a lifetime of inactivity, and that we shouldn't be discouraging people from participating in exercise. I concur that the dangers of inactivity are extensive and widely acknowledged, whereas it's likely that only a small proportion of susceptible runners exhibit chronic maladaptations with ultra-endurance exercise. But my critics have misrepresented the statement as a false dichotomy. There's no expectation that ultra-marathon runners should abandon their sport and assume a sedentary lifestyle, it's just that the long-term risks are likely higher in ultra-marathon than in less extreme behaviors. You don't have to choose between a sedentary life or contesting a series of 100-mile footraces; there's manifest middle-ground, a green zone, which likely provides the best of both worlds.

3.2.9. The Appeal to Anecdote

Aka. misleading vividness. An anecdote is a description or short narrative of an individual's experience, and it can bias you to make hasty generalizations about a product. The aforementioned example of acupressure and the *post hoc* fallacy is founded on a single testimony and is, therefore, anecdotal. Anecdotes are powerful and convincing because they appeal to our emotions regarding everyday experiences; i.e., they're personalized, more so if stemming from someone with whom we have an affinity. But the perspectives of a sole individual can be influenced by any number of factors. For example, they might be emotionally or financially invested in the product, they might have an engrained bias or expectation based on the placebo effect, or their upbringing might predispose them to particular treatments, as might their political, cultural, or religious ideology. These factors can render one blind to a product's true efficacy.

An anecdote is the online review that one reads before buying a product, and no doubt we'd all concur that it'd be irresponsible to base a decision on a single response. Instead, you'd look at the entirety of the reviews and the overall (average) rating, since the mean result is less likely to be influenced by isolated favoritism. You'd be intuitively cautious of a product with only one or two reviews. What might be the critical threshold before you'd note the average rating? Is two, five, or ten sufficient on which to base a decision? More is certainly favorable in this respect, but the review process itself could be biased by external factors: (1) people leaving a review are self-selecting as those types of people to leave reviews in the first instance; (2) people may be more likely to leave a negative online review than a positive one; and (3) those people leaving reviews might be subject to an *investment bias*, whereby they look more favorably on a product after having spent a great deal of money. This further exemplifies the importance of objective evidence, gleamed from controlled scientific studies. Although individual accounts serve some utility in our understanding of a phenomenon, the plural of anecdote is *anecdotes*, not *data*.

The invocation of testimonials ahead of independent scientific research is a major red-flag in product advertising (see the next chapter for a discussion of red-flags), but is a frequent occurrence. Adverts commonly exhibit a *before and after* photo, showcasing a dramatic alteration in someone's physical appearance following usage of the product. The photo has often been doctored in some way, and authentic physical changes take many months or years to materialize, as opposed to a few weeks or months as stated. And should we naively accept the dramatic transition at face-value, there's little reason to ascribe the full effect to the product alone; indeed, there may have been multiple interventions seamlessly integrated. The contrast pictures are powerful ploys because they're visceral, accessible, show dramatic changes apparently with minimal expense, and those changes manifest in someone *real* with whom we can identify, not an abstract celebrity or athlete.

Flawed logic in the form of a fallacy brings into question the validity of a claim. Moreover, when you note the frequency with which these fallacies are committed, both in formal arguments and in the marketing of products, it's clear why objective evidence for a claim is paramount. Principles of logic are violated deliberately or inadvertently; when violated deliberately, to deceive, it's birthed from intellectual dishonesty. When principles of logic are violated inadvertently, it's the result of lack of awareness. Accordingly, when someone commits a logical fallacy in selling a product or position, they're either dishonest or naïve. In either scenario, I'd be reluctant to lend them my custom. And if one opts to buy a product founded on one of these fallacies, based on popularity, clever rhetoric, or a celebrity endorsement, then they must accept responsibility for their actions, especially if the product proves to be ineffective or harmful.

To conclude and reiterate, we must judge products on merit alone, and eliminate the biases evoked by flawed logic. In this chapter, the focus has been on how to action good reasoning in our decisions about product effectiveness, and how this may help us navigate some of the myths and fallacies of the health and fitness industry. We're now well-positioned to more closely scrutinize the *factual* basis for a claim, including a further exploration of scientific plausibility and evidence.

4

SHOW ME THE RESEARCH

ob·jec·tive.

Not influenced by personal feelings, interpretations, or prejudice. Based on facts; unbiased.

4.1. Raise Your Standards

If you've assimilated the discussions of the preceding chapters, you'll know that evidence, clear and objective evidence, is the principal means of discerning if a product is worth your time. The waters become muddied when acknowledging that different systems and organizations have different standards of evidence. For example, that which would stand up in a court-of-law – before condemning a man to spend the rest of his life behind bars – is arguably the most stringent, or at least it should be. The data must prove guilt beyond any reasonable doubt. We've already mentioned the importance of research in a clinical setting. Pharmaceutical companies must provide evidence before new drugs can be sold, and such evidence manifests as robust peer-reviewed studies (randomized-controlled trials) reporting safety and effectiveness. Why should the standards for a sports product be any weaker? This is, after all, a multi-billion-dollar worldwide industry, the products can be expensive, and some people shape their entire lives around the lifestyles that these products and practices engender. More importantly, there's a distinct overlap between products used for performance/health, and those used for health/disease; once you compromise or concede standards in one domain, it's a slippery slope to lowered standards all around. *Pseudo*science breeds *pseudo*science, and tolerating an ethos of poor academic standards has implications for all facets of society.

In most scenarios (whether you're purchasing an ergogenic aid or trying a new training program), there are enough data to make a coherent argument on its efficacy. Published (peer-reviewed) evidence, therefore, should be the litmus test of product efficacy, and we must demand greater emphasis on published studies to support advertising claims. But evaluating research is just one step in determining product efficacy, and there are several others that can predicate your literature search. First, you could look at the advertising claims themselves and determine if they trigger any red-flags, i.e., anything that strikes you as suspicious, incongruent, or that might denote a conflict of interest. Second, you can consider the extent to which the claims are plausible based on our present understanding of science. Finally, the scientific evidence must be evaluated, and we will cover the specifics of reading a paper. While the guidance is primarily for laypeople (those less familiar with scientific research), I'm confident that seasoned academics will also find the discussions of interest. Collectively, the following four steps can serve as a robust, and holistic starting point on critiquing a given product or practice. It's my hope that this guidance will offer some structure and direction in this complex and convoluted domain.

4.2. Step 1: Fun with (Red) Flags

In this business of health and fitness, it'll serve you well to remember that *it's all about the money*. I'm not suggesting that the guy from your local supplement store, or the new personal trainer at the gym, isn't concerned with your fitness goals, but merely charging for a service contaminates their motive. When deciding on the validity of the specific claims, you must take steps to strip away the hyperbole and the sales rhetoric, and give yourself the best chance of being able to accept or reject the product on merit rather than bias.

Your critical faculties are analogous to a Spider Sense, and if the comic-book reference is beyond you, then try Carl Sagan's Baloney-Detection kit mentioned previously. We all have innate aptitudes for detecting things that don't quite fit; we notice when something is out-of-sequence with the rules and systems in which we live, even though we may not be able to articulate or quantify the specific incongruity. For these instances, we employ red-flags, which are a sign of a *fault* or *infraction*. When a product advert or sales-pitch strikes you as suspicious or disingenuous, it's inevitably raised as a red-flag. Too many red-flags are like baseball pitches, three strikes and you're out. Although sometimes our critical-thinking faculties fail – leading us to an erroneous conclusion – we can minimize the risks of inadvertently being deceived, so long as we follow a robust process of analysis. This is as true of decisions in everyday life as it is in the matter of sports science. Below is a summary of occurrences that constitute red-flags.

Claims that commit logical fallacies. The aim of the last chapter was two-fold: (1) to help you identify when a logical fallacy is being committed in the *sale* of a product; and (2) to help you identify when you might be committing one of these fallacies in your own arguments. For example, is there an attempt in the marketing rhetoric to blind you with science, using technical jargon to lend false legitimacy to the product? Do the advertisers use ambiguous terms like *detox*, *wellness*, and *vitality* which don't have concise definitions and are, therefore, easy to defend when challenged? Does the sales pitch invoke words like *natural* and alternative? We must strive to expose those claims that don't speak directly to the efficacy of the product. Fallacies are smoke-screens designed to confuse and deceive, and while we're distracted by a fallacious appeal to tradition, nature, or popularity, the actual evidence remains ill-considered. The first potential red-flag, therefore, is if the product is sold on the basis of a logical fallacy.

Products sold on testimonials and anecdote. Testimonials are often provided in-lieu of credible (objective) evidence. Aside from the online review, the larger sports companies have sufficient capital to pay sportsmen and sportswomen to model their garments, run in their shoes, and be pictured drinking their supplements and eating their vegetarian sausages (see the fallacious *appeal to authority*). In many such adverts, the athlete will imply that their phenomenal success, at least in part, was attributable to the product and its performance-enhancing effects. As an athlete, elite status remains well beyond my grasp, but I've nevertheless competed in a few high-profile extreme running races, and sports companies have gifted me product in return for a personalized endorsement. Such sponsorship can be career-suicide for a professional scientist, unless one has already attained a certain lofty level of success such that nobody will question your motives. From a scientific perspective, anecdotes from a celebrity or Olympic athlete are no more meaningful than that from a user of your local gym, because neither one speaks to the objective workings of the product. Consider the sheer number of people who would be willing to offer an endorsement in exchange for money or free goods. Accordingly, even the most heartfelt of testimonies cannot be guaranteed, and testimonials in place of data reflect another red-flag.

Products sold as alternatives. In Chapter 8, we plunge into the controversial world of complementary and alternative medicine. Products in this domain are often marketed as natural or traditional alternatives to modern science and, in the process, claims sometimes demonize the latter in an attempt to carve out a sales niche that cannot be scrutinized by the traditional scientific method. Remember, modern science and medicine operate on rules and regulations and stringent evidence thresholds; there are no such systems that restrict the sale of alternatives. Evidently, you're much more likely to be deceived by something marketed as such.

Products sold with emotive language. A claim invoking words like *breakthrough*, *miracle*, or *secret* is appealing to our emotions. But these aren't terms you'll ever see printed in a scientific paper because, by definition, they're not scientific. Consider the clickbait that appears in the side-panels of online articles and search engines; they usually feature terminology like this, designed to bypass your critical faculties and exploit your impressionable *chimp brain*.

Products offering rapid results. Nothing comes for free; no more is that true than in health and fitness. Certainly, there are some interventions that work faster than others, and select supplements can confer an immediate outcome, but most highly valued physiological adaptations must be toiled after. Endurance, speed, strength, power, skill, decreases in body fat, bigger muscles, and washboard abs take months or years to develop (what's more, they're quickly lost). For the most part, it's a red-flag when products offer dramatic outcomes in an unreasonably short time-period.

Products making multiple claims. In the same manner as that of traditional Snake Oil, many interventions are marketed as *cure-alls*. These items are accompanied by a series of claims usually about unrelated phenomena. If the product works via a clear mechanism of action, then it's unusual for it to influence multiple different body systems. For example, it's notoriously challenging to concurrently build muscle and lose fat. The former requires a positive nitrogen balance, which rarely occurs unless the organism is in a state of matched or surplus caloric intake. By contrast, a prerequisite for losing fat is a negative calorie balance (i.e., burning more calories than that consumed). There are very few instances that arise in which both processes occur simultaneously. Supplements should be regarded cautiously, therefore, when they claim to simultaneously and directly bestow its user with both. Increased muscle mass, decreased fat mass, enhanced endurance, flexibility, and strength; these are all governed by different processes. Be wary of products or practices that claim the capacity to endow you with the lot.

4.3. Step 2: Prior Plausibility

If you were going to make a large cash investment to conduct a test, or series of tests, on the effectiveness of a new intervention, wouldn't you first make an educated guess as to how likely it was that you'd obtain meaningful results from your investigation? It'd be a shame for all that money (not to mention time) to be squandered investigating something inert. Scientists sometimes get their research back-to-front, because a great many studies have been conducted on products and procedures which one could have guessed in advance weren't going to be effective. This is what's meant by *prior plausibility*. When asking if something has prior plausibility, we're essentially posing the following question: *given our current understanding of nature and how things work, how likely is it that the claim(s) being made about this product are actually true?* Although this step requires an understanding of the relevant science (biology, for example), this can be acquired to an extent, but that's also how we should use subject specialists. Another way of considering prior plausibility is in considering how *realistic* it is that the product will function in the manner proposed. Rarely can a potential outcome be determined as a binary plausible/implausible; instead, it's a continuum that's anchored at one end by *completely implausible* and at the other by *completely plausible.* Conjecturing on a product's prior plausibility will help pre-empt your comprehensive investigation.

It's unethical to devote resources to a project which is unlikely to yield meaningful results. Note that I employed the term *meaningful* rather than important; only a minority of studies submit important or breakthrough findings, and these are predicated by a series of step-wise increments in the research, mostly comprising data altogether less fantastic. But results that are *meaningful* are those that are valid and reproducible, assessing real phenomena, that can be tested or quantified in some way. Not only are funding and lab time valuable commodities, but data must be meaningful to justify risking the welfare of individuals who might volunteer for the research. When my undergraduate students propose ideas for their final-year dissertation projects, we review their ethics applications for health and safety risk assessments, and scientific rigor. Their documentation may be exemplary, with carefully considered inclusion criteria, forethought of the risks associated with participation, and any emergency procedures that might be necessary. But if the study itself is unlikely to yield any worthwhile results – e.g., because they're proposing to test an alternative remedy that's been fully disproven – then it's ethically unsound to recruit subjects for the study, particularly if the procedures involve exercise, blood sampling, or other activities with inherent risks. Accordingly, giving thought to prior plausibility is paramount in advance of any decisions that might be made regarding an intervention.

There are three basic scenarios that might arise relating to prior plausibility. First, there are products with a reasonable mechanism of action, but no available evidence. Second, there could be a product with some interesting preliminary data suggesting efficacy, but no known mechanism of action; i.e., according to some reports, it seems to work, but none can provide a lucid explanation as to *how.* In both instances, one could argue that bigger and more well-controlled studies might yield a deeper understanding of the product's worth. The third scenario encompasses products that have no coherent mechanism of action (i.e., that which is proposed doesn't agree with our current understanding), but that also have no available or supporting evidence, even preliminary data. The academic outlook is unfavorable for products that fall within this category, and there's very little incentive to regard any associated claims. Some products entirely break the mold by not only being inherently implausible, but impossible. For these to influence physiological function in the way described, we'd have to rethink the laws of physics, chemistry, and biology. There are several sports science products that fit into this category which we will consider shortly. Such products not only lack a known mechanism of action, but they also violate basic laws of science. Physicians and scientists, for example, are in general agreement on the plausibility of homeopathy; an alternative medicine in which patients are treated with ingredients of such minute concentrations, the preparations are indistinguishable from water. From www.sciencebasedmedicine.org:

> Homeopathy violates the law of mass action (a basic principle of chemistry), the laws of thermodynamics (extreme dilutions maintaining the chemical "memory" of other substances), and all of our notions of bioavailability and pharmacokinetics. Homeopaths, therefore, substitute any notion of chemical activity with a vague claim about "energy" – but this just puts homeopathy in the category of energy medicine, which is just as implausible. Invoking an unknown fundamental energy of the universe is not a trivial assumption. Centuries of study have failed to discover such energy, and our models of biology and physiology have made such notions unnecessary, resulting in the discarding of "life energy" as a scientific idea over a century ago.

An important caveat of note is that regardless of the prior plausibility (or lack thereof), the available data should always be considered independently. Data can, on occasion, be ignored, if obtained via a flawed study with methodological insufficiencies, in which case such findings offer few lucid insights. But that obtained from robust, well-controlled studies will always be of greater value. A lack of prior plausibility, therefore, isn't justification to disregard scientific findings; it just means that *extraordinary claims require extraordinary evidence*. Similarly, a product that is *completely plausible* may have little applied value because some minor detail pertaining to its function has been overlooked, or a new discovery has rendered the premise void, or logistical reasons render it impractical (e.g., it'd be too expensive to manufacture on a large scale). The scientific process by which outcomes are tested – when conducted with sufficient rigor – is designed to extol that which works, and abandon that which doesn't. In this way, good science ensures continued forward motion in the acquiring of new skills and interventions.

Some scenarios confer complex oddities in plausibility and evidence. As a brief thought-experiment, let's consider the much-practiced strategy of simulated altitude training. Once a purely scientific endeavor, tested in modified physiology labs and altitude chambers, it's since been heavily commercialized and well-marketed. Altitude is now a trendy intervention aimed at fitness, fat-burning, and immune function. Several companies specialize in altitude training for athletes and the general population that want to *improve speed and endurance, strength and power, energy and wellness*. The proposed mechanism of most simulated altitude training involves artificially reducing the oxygen content of air in order to stimulate the bodily production of red blood cells (RBCs); this is referred to as normobaric hypoxia, because even though the oxygen content has been reduced (hypoxia), the barometric pressure has remained (normobaric). Since RBCs are principally responsible for the transport of oxygen around the body, such an adaptation would increase the capacity to deliver oxygen to the exercising muscles, thereby improving aerobic endurance. The premise, to artificially stimulate RBCs via this method, is sound; i.e., living or training in such an environment is known to have a physical affect, and there's plenty of supporting literature. Nevertheless, the research pertaining to simulated hypoxia is inconclusive with a great many variables to consider, including the duration and frequency of exposure, the oxygen percentage of the inspired air, the training intensity required to induce optimal adaptations, the nature of the sport being contested, and human non-responders, all of which mediate the effectiveness of the strategy. Furthermore, performance in sports that emphasize maximal capacities and explosiveness (e.g., short-distance time-trialing, martial arts) may be unaffected or even compromised because the time spent exercising at very high intensities is diminished; altitude interventions may not be worthwhile for all sports. While there's plausibility to the mechanism, the benefits of altitude training are far from assured. So, when it comes to selling altitude training as an intervention, the logical premise and plausible mechanism are alone insufficient factors on which to hang your hat. For this – as for every other product and intervention – we must demand something more from the manufacturer.

4.4. Step 3: Show me the Research!

And so, we arrive at what for many health and fitness products will be the end-of-the-line: an analysis of the scientific evidence. It would be my preference that everyone had equal education in critically appraising evidence, and have access to the same high-quality, unbiased information. In reality, this isn't so. Information is biased, not of equal validity, not equally distributed, or freely available, nor is education, inclination, opportunity, or resources. This is decidedly problematic when one's understanding of a product and its worth are dependent on these factors. In terms of reviewing scientific evidence, many science websites – several independent of education and government – exist to do the heavy-lifting on your behalf, providing concise and accessible summaries of key topics in health: Science Based Medicine, Snopes, and IFL Science to name a few. Websites providing this service specifically for sports products are relatively scarce. But a responsible consumer, and any true critical-thinker, should have the capacity to review literature for themselves. Accordingly, this section provides an abridged overview of the scientific publication process, and some of the important considerations when determining product efficacy in this manner. Succeeding it are the specifics of how to read and critique a journal article.

In order for a company to cite published studies to support a claim, or for review articles to summarize the research, it necessitates that those studies have first been conducted. Often, there's a dearth of literature for a product that's already on sale, for one of two reasons; first, studies on the effectiveness of an intervention have not been conducted by the company for fear of unearthing negative results, thereby exposing the product as a sham. Second, there's no independent research (from labs not affiliated with the manufacturer) because scientists consider the intervention to have poor prior plausibility and, therefore, not worth the investment. There's an enormous wealth of literature on all manner of topics, and many commercial sports products have at least some manifestation of associated data, regardless of the quality. It speaks volumes, therefore, when a product has no associated data. Consider what it means to be *published*; we're not referring to articles published in mainstream health and fitness magazines alongside adverts and press-releases. And we're not referring to a glossy health supplement dispensed free with your Sunday paper. We aren't interested in blog posts, or something written in a book. By *published*, we mean in a peer-reviewed scientific/academic journal. It's a distinction of paramount importance; writers can operate unrestrained in magazines and websites, jotting untruths, misinformation, and opinions at their discretion, either deliberately or inadvertently. Peer-reviewed papers, however, must be subjected to (and graduate from) a superior level of academic scrutiny. This independent substantiation is the hallmark of scientific self-policing.

Conceiving hypotheses. Recall that a hypothesis is a proposition that one might use to try and explain a phenomenon. It's an evidence-based, educated guess that offers a foundation for further investigation. A hypothesis, therefore, is distinct from a theory. Scientific research often begins with a hypothesis on the outcomes of a given intervention, or the expected responses in an observational study. The scientific process hinges on our ability to formulate hypotheses, and then design appropriate methodologies to rigorously test them. Once the statistics have been performed and the results collated, the hypothesis can be accepted or rejected based on the findings. Given that the aim of experimental studies – those testing the effects of an intervention – is to test a hypothesis, the Introduction of a paper should end with at least one hypothesis.

Designing a study. Ideally, a product idea will have prior plausibility and a clear hypothesis of postulated effects. A scientific hypothesis can graduate to a comprehensive scientific *theory* only when robust data are available from multiple different lines of evidence to corroborate it. Recall that the term *theory* in science is more-or-less the greatest accolade attainable by a system of thought, and a *law* is that which can be described comprehensively using mathematical equations (e.g., the laws of thermodynamics); everything else is restricted to a theory. But theories begin life as hypotheses that require corroboration. For example, imagine I design a brand of running shoe and hypothesize it to reduce injury prevalence in marathon runners by decreasing impact forces transmitted through the lower-limbs; my invention is predicated on a new sole that absorbs more impact than the regular shoe. I can then design a study, or series of studies, to test my hypothesis. If I establish no difference in injury prevalence, I should discard my idea (reject my hypothesis) and move onto the next, assuming the study is sufficiently rigorous that I'm confident in the results. But if the findings permit me to accept my hypothesis, it justifies further research and independent replication. The precept is that products should be conceived and then rigorously tested *before* going on sale – just like drugs and other medicines – whereas a great many products are presently available, with manufacturers contriving to retroactively fill the void of published studies. Conducting science this way contaminates the process; it's a backward system. What's more, if such *post-hoc* research shows no evidence-for-efficacy, it can be ignored or discarded prior to publication. In such a system, it's axiomatic that convincing evidence-for-efficacy is not a prerequisite for product sales, and this is what we must strive to overturn.

The peer-review process. For the non-scientist, let's take a moment to turn transparent the black-box of peer-review. The basic process is simple enough. Scientists at an institution (a hospital, a company, a university) conduct some experimental research on the effectiveness of an intervention. The data are analyzed, and the findings drafted in a lab report. These reports follow a standard structure referred to as IMRAD, comprising an Introduction to the topic, a description of the Methods utilized, the detailed Results including relevant statistical analyses, and a general Discussion in which the results are interpreted, compared to previous literature, and an attempt made at an explanation. The final manuscript, once agreed upon by all corresponding authors, is submitted to a relevant journal. There are well over 100,000 different academic journals publishing six million scientific papers each year, in disciplines and sub-disciplines from physiology, to sociology, to physics. Once submitted, the journal editors appoint several experts in the field to review the paper (for free), and offer critical feedback on the rigorousness of the methods and validity of the conclusions. Often, a paper is returned to the authors with comments (sometimes minor and other times major), and those comments are addressed before the manuscript is resubmitted for a second round of reviews. This process may be repeated for many months until all parties mutually agree on the finished report, at which point the paper is published and incarcerated behind a pay wall. If the first review is particularly unfavorable, the paper is rejected on the grounds that it hasn't surpassed a minimum standard, or perhaps isn't the right fit for the journal. The paper may also be rejected before being commissioned for review. Generally, the more rigorous the science and important/impactful the findings, the higher the quality of journal in which the paper will be published. The system described is the best we have at ensuring quality and rigor in research. Unfortunately, most people engage more with the simple and superficial testimonies of journalists or athletes reporting in the media on a sensational and miraculous new training strategy, or the incredible weight-loss achieved with a new diet pill, than the detailed and sometimes laborious explanations of a dreary scientist residing priggishly in a lab. But critical is the peer-review process; it's a valuable self-policing system whereby scientists attempt to maintain a minimum standard of practice in their colleagues, and it's all we have to authentically distinguish accurate from inaccurate claims.

Not all papers are created equal. Peer-review is imagined to be a transparent process, emancipated from mortal flaws like bias and subjectivity, but the system is far from infallible. There are well over 200 journals in the disciplines of physiology and nutrition alone, all ranked according to certain metrics, including the impact factor and journal rank, details not paramount here. Some are revered and respected, with a reputation for high-quality research, and others less so. Irrespective of the standard of work, it can always find a home; indeed, many journals feature studies of a decidedly dismal standard. It's the responsibility of those reading the paper, however, to interpret the findings on merit, and reason upon its quality. Some journals publish nominal case-reports of inconsequential data, while others insist only on high-impact and robust observations. There are online lists of journals devoted to every topic, including alternative remedies and *pseudo*science which are likely more sympathetic (and less objective) toward studies that reinforce their narrative. One should, therefore, always be cognizant of the journal in which the report is published. It's not a definitive indication of quality, but it's a start. Journals like *Science* and *Nature* are generally known to be highly reputable, whereas the *Albanian Journal of Alternative Medicine* is probably less so. Be wary of questionable journals or those from natural/alternative organizations with very low publication standards, many of which appear to be authentic peer-review outlets but aren't (see *Predatory journals*).

Several other issues deserve consideration. The *peers* tasked with conducting the reviews are academics like me; they're often developing manuscripts of their own, or laboring to finalize a grant application, or a book chapter, with cascading teaching and administrative duties. Time is a valuable commodity, and rarely is there enough to permit a meticulous review of every paper they're sent. Accordingly, the thoroughness with which reviews are conducted is highly variable. It's due to fallibilities like these that some alternative therapy proponents propose a complete boycott of the peer-review system, i.e., we should abandon peer-review and permit the public to arrive at their own conclusions. But any rigorous scrutiny of evidence is likely to be harmful to proponents of alternative therapies. The studies that show their treatments to be effective are often of a lower standard, and discarding peer-review would serve their cause tremendously. Ours is not a perfect system, but it's a damn-site better than what came before, and much preferred to the bedlam you'd see in its place. A broken system is more effective than no system.

Occasionally in contemporary culture, scientists promote their discoveries directly through the press, bypassing the review process. This is often to conceal insufficiencies in the data that would otherwise be exposed by peer-review, such as low subject numbers, fabricated data, inappropriate statistical analyses, etc.

Envision a world in which this was the norm. On the value of basic scientific research, the International Council for Science says:

> Basic scientific research is defined as fundamental theoretical or experimental investigative research to advance knowledge without a specifically envisaged or immediately practical application. It is the quest for new knowledge and the exploration of the unknown. New scientific knowledge is essential not only for fostering innovation and promoting economic development, but also for informing good policy development, and as a sound foundation for education and training.

Predatory journals. Each day I'm afflicted with at least five emails from predatory journals, each one brimming with lavish praise, gushing at how much of a distinguished honor it would be to publish my work. Fewer than one in ten is even within my research domain. Predatory journals are phishing scams; every email address on which they can lay their hands gets spammed with fake journal metrics and calls for Abstracts, quick-to-publish journal submissions, and conference presentations, all for a fee. As with the more generic hoax, the emails are targeted at the lowest common denominator, to immediately discount all but the naïve and the desperate. The process to which a paper would be subjected is scarcely peer-review and, if it is reviewed, it's a perfunctory process of *review* in name, only. In one recent analysis of predatory journal activity in medicine, the mean author charge per article was US$634.50, and when the office locations of those journals were scrutinized, they were often found to be unreliable (e.g., supermarkets, highways, football fields, postal boxes, etc.)[12]. Those who publish in predatory journals tend to be young and inexperienced researchers from developing countries, and can be distinguished by their economic and sociocultural conditions[13]. Correspondence from a predatory journal can be discerned by its crude attempt at formality and flattery, with emails often composed in broken English addressed to *distinguished professor*, titled *Greetings of the day.* These journals are pay-to-publish scams, and while online databases like PubMed do a commendable job of sifting them out, some inevitably slip through. Comprehensive lists of them can be found online.

Cherry-picking. Occasionally, proponents of a product acknowledge the need to support their scientific claims with evidence, and this is a noteworthy progression in enhancing the integrity of the industry. However, in the search for that illusive scientific credibility, they'll often try to shortcut the process by carefully selecting studies that support their claims, while deliberately omitting those which don't. This strategy is known as *cherry-picking*, and it's widespread in commercial science. For example, in my research for this book, I contacted several companies and manufacturers of various products with a request for data to corroborate their claims. But in return, rarely was I supplied with anything that accurately reflected the available consensus. Cherry-picked research showcases only one side of the argument, i.e., the side that supports the assertion, while simultaneously ignoring that which doesn't.

Review articles and meta-analyses. If you're without the skills or resources to assimilate countless papers, sometimes in an unfamiliar field, then a literature review might be preferable. The authors of these helpful summary articles have laboriously read and summarized the studies and drawn conclusions on the current understanding. They're sometimes called survey articles. Meta-analyses are discerned by their statistical incline, designed to collate data from a number of studies, including reviews. Depending on the extent to which you want to *understand* a topic, you'll then need to perform a deep-dive on the literature by seeking out individual papers and more closely scrutinizing the Methods and Results. An online search for *reviews* will likely return a variety of types, and it's worth briefly noting the differences among them. A systematic review will synthesize and assess the existing studies with the aim of answering a specific question, following a strict and transparent method to minimize bias. By contrast, narrative reviews will describe and discuss the science, but without explicitly stating the means of gathering and presenting evidence. The latter is more subject to bias because the literature can be shaped to provide a *narrative*, and the quality of the studies included isn't always explicit. A narrative review, therefore, might provide its conclusions without necessarily indicating poor study design or conflicts of interest. Narrative reviews can be more comprehensive, however, since they're able to cover a wide range of issues without the restrictions of systematic analysis. Meta-analyses, as mentioned, will invoke comprehensive statistics and disclose details of systematic bias in the literature. In any case, collating reviews can be an excellent means of familiarizing yourself with a topic, provided you take time to understand their inherent limitations, and look more closely at the studies they cite, when relevant.

Publication bias. There's tremendous pressure on both scientists and journals to publish novel, impactful research. On more than one occasion, I've conducted an experimental study, written a clear, concise, and highly polished manuscript with robust methods and data analysis, only to see it outright rejected from a journal on the basis that the intervention didn't evoke any statistically significant responses. These kinds of studies (with null findings that don't necessarily support the hypotheses) play an important part in our overall understanding of a subject, help shape the future direction of research, and discount or discredit existing practices. Yet, some journals are reluctant to publish reports with null findings because they're not considered novel when, in fact, they may be quite the contrary.

The *AllTrials* campaign (www.alltrials.net) was launched in January, 2013, in response to widespread frustration with the non-reporting of important clinical data. The Declaration of Helsinki – the World Medical Association's statement of principles for medical research involving humans – decrees that all clinical trials must be registered, and the findings transparently reported irrespective of the study outcomes. Presently, it's the researchers who dictate what data are, and aren't, reported. Their decision may be influenced by the study findings, e.g., choosing not to publish null findings for fear of contradicting the existing trend. This, in turn, leads to publication bias in the literature. The evidence suggests that approximately half of all clinical trials have never been published, and that trials with negative outcomes about a treatment are less likely to be submitted for publication[14]. The *AllTrials* campaign calls for all past and present clinical trials (experiments or observations in clinical research) to be registered, and their results reported. Since its conception, the *AllTrials* petition has been signed by nearly 100,000 people and nearly 800 organizations. These heartening initiatives are critical for restoring transparency to the publication process. Practitioners and regulating bodies must have access to the full, comprehensive results of studies in order to make informed decisions about interventions. We should obligate researchers to implement robust procedures and report the findings irrespective of their bearing; in the long term, it will lead to a more thorough and legitimate understanding of a field. Imagine the changing landscape of sports products and the wider health and fitness industry if there were a legal obligation to publish studies with null findings; the conclusions surrounding countless products and practices would be rapidly reconsidered.

Students frequently make null observations in their experimental studies – usually they're testing the effects of a supplement or other ergogenic aid because the study design is simple – and they'll fervently search for any number of excuses and explanations for why they failed to observe a significant effect. The studies are frequently under-powered (too few participants), or there are basic errors in the precision of the measurement techniques which increases the variability. They'll invoke other excuses pertaining to supplement non-responders. But rarely do they have the confidence to conclude that the thing just *might not work*. They're primed to interpret a null response as erroneous, less important, or less valid, even to the extent of querying if non-significant findings will influence their grade. What a question to ask. We coach them, as best we can, to acknowledge that science is a *process*; it's on the research question, clarity of the rationale, study design, and rigorousness of the methods on which they're assessed. Assuming the conclusions are valid, and supported by the findings, the outcome is irrelevant. But it's a hard pill to swallow when academic journals sometimes don't embody the same ethos, showing favoritism to studies with positive findings. We mustn't let deficiencies in the system dictate our study design or manuscript preparation; scientists must exhibit the courage and academic integrity to prioritize the process, and accept any and all conclusions that result.

Conflicts of interest. In assessing the validity of a piece of research, we also need to consider if there are any conflicts of interest; e.g., are the studies funded by industry, or is the lead scientist working for the manufacturer? This isn't catastrophic for a study but it could be considered a red-flag. Occasionally, we see a single lab publishing the predominance of data in support of a supplement, with others struggling to replicate the findings. Ideally, the studies would be repeated by third parties who observe the same findings using the same (and sometimes different) techniques. These additional steps are important to endow the product with additional scientific credibility, and afford us confidence in its efficacy.

Replication of studies. The current system places too much emphasis on originality in research. It's important to make a novel contribution and push the boundaries of human knowledge, but researchers often prioritize this above all else; unsurprisingly, given that journals prefer publishing papers with unique findings. But replication – that is, repetition of a study – is crucial in order to determine if the basic findings of the original research are valid and authentic. Replication can make use of an identical approach (a similar population and methods) to assess reproducibility, or different participants and circumstances to determine if the findings are more broadly applicable. One study is scarcely sufficient on which to base public understanding.

4.5. Step 4: How to Read a Paper

People read published research papers for several reasons, including preparing for a class, preparing for a conference, staying current in a field, and to establish the efficacy of certain products and interventions. While career scientists might spend hundreds of hours each year deliberating articles, many university-level science courses demand such familiarity from students without providing any formal instruction on the process. Consequently, students learn by trial-and-error, making numerous interpretive mistakes and squandering much time and effort. Many scientists *still* haven't acquired an effective means of analyzing a scientific paper, but it's a skill like any other. I hasten to add that reading papers isn't just for scientists; non-scientists cannot rely on third-party news reports or those which they read on social media. It may be extremely useful to learn the skills for yourself.

A cursory study of the Abstract is acceptable to determine if the paper is of interest (i.e., if it's relevant to your current search), but it's wholly insufficient to discern the nuances of the method or any flawed approaches, since these can be justifiably omitted from an Abstract of limited word count. Regardless of the discipline, there are some common strategies you can employ to help you assimilate the important information, while simultaneously noting the strengths and weaknesses of the study so that you can establish to what extent the findings are robust. There are two brief caveats to note before we start. First, the forthcoming guidance isn't intended to patronize any seasoned academics who may be reading; it's a brief summary aimed primarily at laypeople, although I'm sure there are several precepts useful to the scientist. Second, this isn't a comprehensive guide. There are complex minutiae to reading and fully assimilating a paper – particularly with respect to sample sizes, reliability, and interpreting statistical analyses – for which there's insufficient scope to discuss here. Moreover, a full exploration of technical statistical terms will only detract from the major theme of the book (which isn't statistics). Still, the ability to fully interpret the findings of a paper depends to a large extent on an appreciation of statistical modeling; given that there are entire textbooks devoted to these topics, I encourage further study (especially for students and academics).

Finally, recognize that scientific papers once replaced personal letters as the primary means of communicating new scientific discoveries among natural philosophers; they were never intended for disseminating new findings to laypeople. They also, by design, comprise sufficient detail to allow other scientists to fully replicate the study. Reading an article in a journal, therefore, is a very different challenge to reading one in a magazine or on a popular website, not just because the former is usually longer, but because you don't read the various sections sequentially, from start-to-finish. Throughout, you'll also need to remain cognizant of the factors aforementioned including the quality of the journal, the obvious conflicts of interest, publication date, etc. As with any skill, proficiency comes with time and practice, but the benefits will more than justify the effort.

Getting access. First, you'll need access to the article in question. Google Scholar is an accessible and familiar search engine, and most scientists/medics will use an online database like MEDLINE or PubMed. They all function on the same basic principle of key-word searches, so use several that sufficiently characterize the topic in question. If you belong to a university or other institution with academic credentials, you'll have access to most articles; if you don't, there are several options. You can filter your search (in Google Scholar, for example) to return only those papers available via open-access (i.e., with an open license that removes use and reuse restrictions). You could, of course, pay for the journal subscription and/or the cost of the individual article, but these are not reasonably priced. It's likely you'll know someone in your close personal or social media network with academic credentials who could access papers on your behalf. You can try joining ResearchGate – a social networking site primarily for scientists and researchers to share papers – but you can often get access with a non-institutional email address. You can also try contacting the author directly with a polite email asking for the PDF; in most cases, they'll be happy to oblige, and their contact details will be on ResearchGate as well as on the front page of the journal article under *corresponding author.* Finally, if all else fails, the catalogues of many large libraries (actual physical libraries…) comprise books, journals, and manuscripts (as well as maps, stamps, music, audio, patents, photographs, and newspapers). On many occasions over the years, I've visited the British Library near King's Cross station, London, to access a specific journal article that was unavailable anywhere else; and I'd encourage the trip if you're close-by. I've also visited the New York Public Library which has a richer history, but both journeys were equally inspiring.

The Introduction; don't start at the start. The Abstract offers a concise overview of the study's procedures and main findings. It's often written to *sell* the study (to the reviewers and journal editor) and emphasize novelty to aid in publication. Indeed, when papers are written, it's assumed that readers will begin with the Abstract. They're also word-limited, and so may omit important details that would otherwise be needed for a clear interpretation of the conclusions. For these reasons, it's important that you *don't* begin by reading the Abstract. The Introduction, by contrast, provides basic background information on the area. For a non-expert, it's perfect for updating your existing understanding of the topic. It should rationalize the study, and frame it in the context of previous research. The Introduction should also conclude with a short sentence on the study aim(s) and research question, which will predicate your assessment of the Methods. Skim-read the paper first, and note any words with which you're unfamiliar so that you can study them later. Acronyms are common in scientific papers, and should always be explicitly stated at first use. The Introduction is written in a charitable way; it leads you carefully by the hand, navigating through the twists and turns of the existing research narrative, before unceremoniously dropping you into the deep, dark methodological abyss.

The Methods. Following a statement of the research question, the authors will reveal the approach they adopted to address it. What distinguishes a strong from a weak study is the quality and appropriateness of the Methods. Poor studies are often characterized by ambitious aims with disproportionately unimpressive methods that are insufficient for the task, and this should be at the forefront of your mind during your analysis. The Methods contains a lot of detail – enough to allow a full replication of the study – and it may be impractical to digest in one go. Instead, deconstruct the various sections, read them piecemeal, and consider compiling your understanding of the protocols in a flow chart, spider diagram, or other illustration. Think carefully about the participants used, the sample size, and whether it was appropriate for the analysis. For example, very mechanistic and basic science studies tend to have fewer participants, because it's difficult persuading many people to have catheters inserted or biopsies extracted, whereas large RCTs and drug trials will recruit considerably more. The letter 'n' is used to denote the *number* of participants involved (i.e., $n = 100$). Small samples can sometimes lead to under-powered studies which increase the chances of a false negative (a type II error, or incorrectly calculating a non-statistical significance), particularly if the variance is high, whereas samples that are too large increase the chance of a false positive (a type I error, or erroneously finding statistical significance). To appreciate the nuances of sample size, some additional reading will likely be required. Focus your attention on the specific techniques employed, and the apparatus, and keep in mind that good studies will justify their procedures with appropriate references to reliability and validity testing. Non-subject-specialists may need to be selective with which details they assimilate and which they discard.

As per my earlier comment, interpreting the statistical output will necessitate prior knowledge of the discipline. Presently, it's sufficient to note that there is a finite range of appropriate statistical tests for each study design, and the tests employed must be selected in advance of data analysis.

The Results. The detail in this section will vary greatly among studies. Some psychology papers, for example, report the full statistical model and all associated output, including main and interaction effects, degrees-of-freedom, confidence intervals, p values, and effects sizes. While transparent and technically correct, it's my view that such an approach increases the risk of the overarching message becoming lost among a sea of numbers and symbols. By contrast, many physiology papers make only a perfunctory attempt at basic statistics, by reporting only standard deviations and p values. This makes it impossible to formulate a full understanding of the Results, or make valid inferences from the data; analogous to tearing out random pages from the middle of a spy-novel, resulting in a disjointed and incomplete narrative. The best papers, again in my view, compromise between the two approaches; they're transparent with the statistical approaches employed but provide the detailed output only for key outcomes. The full analysis can always be made available as supplementary data to permit transparency without obscuring the message. When reading the Results, it's crucial that you don't try to *interpret* the findings, although it's normal for your brain to start proposing mechanisms and establishing connections between datasets. Interpretation and explanation are reserved for the Discussion, so for now, test your comprehension by attempting a short written-summary of the key findings. Take special note of whether outcomes are statistically significant or non-significant. Also, closely scrutinize the quality of any graphs presented; check that axes titles are present, the units are correct, the scales are appropriate, and error bars are present (if not, then ask why not?). These factors allude to the overall quality of the report. Some people advise non-experts to read the Discussion before the Results to acquire an overview of the key findings, but whether you do so will be largely dictated by your understanding of the subject, as well as personal preference.

The Discussion. Before reviewing the authors' analysis and interpretation of their data, take a moment to summarize for yourself the implications of their findings. Specifically, consider whether the authors have addressed their original research question, and what the broader implications of the data might be. The extent to which you agree with the arguments in the Discussion will vary, but be cynical; make a concerted attempt to find fault with their study and their assessment of the data. Consider ways the study could have been improved, or methodological insufficiencies that might invalidate the conclusions. Good reports will highlight any *technical considerations* (i.e., limitations) to the study but, if omitted, do this on their behalf. While it may seem petty and pernicious to emphasize the negatives of their approach, science thrives on constructive scrutiny, and this rigorous process engenders a more valid outcome. Read over the conclusions several times and assess whether they're supported by the data, and whether you agree with the authors. This final step is far from redundant; bias is a powerful force and, when combined with the ubiquitous pressure for novelty, can lead to false and misleading conclusions. To close your review, return to the Abstract and compare its message to that represented by the full report. It'll also be constructive to integrate the reviews of colleagues, friends, and other authors, and whether the study represents a scientific consensus or instead an isolated opinion. If it's a large study with a major finding – particularly one that's captured the public's imagination – there will be published responses and rebuttals wherein other experts scrutinize the data more closely.

The Three-Pass Approach. Reading a paper cover-to-cover, assimilating minute details while interpreting the complex statistical output, can sometimes be a long and laborious task. To improve your chances of a productive read, a three-pass approach has been proposed[15] during which each read builds on the last to accomplish a specific set of goals. This strategy is aimed more at scientists, but could be adapted for the non-expert. During the first pass, a scan of the paper is suggested with the aim of acquiring an overview of the important theme(s). Don't spend an inordinate amount of time understanding the nuanced approaches or getting into the *nitty gritty*. Read the Introduction, Conclusions, and the remaining subheadings, only; ignore the rest of the paper. Following the first pass, you should be able to determine the novel contributions of the paper, discern whether it's well-written, and whether the assumptions appear valid. At this juncture, one of three reasons might preclude reading any further: (1) the paper doesn't interest you or wasn't as relevant as first thought; (2) you're insufficiently knowledgeable of the topic to really assimilate the arguments in greater detail; or (3) the author has used inappropriate methods, or makes invalid assumptions and/or conclusions, that limit the utility of the report.

The second pass involves reading the paper with greater care. The background information has been dealt with, so cast aside the Introduction for now. Labor over the detail, perhaps making notes in the tight margins. Read the Discussion and, when prompted, look at the figures and tables sequentially and dissect them accordingly. After the second pass, you should be able to summarize from memory the main crux of the paper. It should only take around one hour to complete the first and second passes combined, but pouring over detail for considerably longer renders the process unsustainable for multiple papers. A third pass is usually reserved for fully understanding the paper and its finer details in order to progress in your field of study, or if you're providing a review for a journal.

Should a third pass be required, you should virtually re-implement the paper, i.e., imagine conducting the study for yourself, using the same study population, methods, and equipment as that stated. Such a perspective allows you to view the study through the lens of the researchers, balancing all of their motives and assumptions, the appropriateness of the measures, and the suitability of the apparatus. Every assumption in every statement should be scrutinized. You might continue your notes and scribbles until the paper resembles a Jackson Pollock: link sections, refer to other papers, and jot ideas for future studies. A further hour may be required for an experienced reader to conclude the third pass, but it may take longer. By the close of your review, you should be able to recall the study (the majority of the details) and recount its strong and weak points to a third party.

4.6. Statistical versus Clinical Significance

When a study draws attention to a statistically significant result, be mindful that it's based on the arbitrary alpha level denoted in the Methods (usually 0.05). If the statistical analysis identified a *p* value of less than 0.05, then the test is considered to be statistically significant; this statistic permits the assumption that there's a less than 5% probability that the result was due to chance. This is a dramatic oversimplification of a very complex area which, again, may need further study. In any case, this arbitrary *p* value says nothing about the importance of the result in the real world, or its relative magnitude. For the latter, we rely on other statistics and metrics like effect sizes. But whether something is *clinically* significant depends on its practical importance. Consequently, a result can be statistically significant but not clinically meaningful.

In a review I published in 2019[16], I collated the studies assessing lung function following marathon or ultra-marathon running. All of the 15 studies reported statistically significant post-race decreases in one or more metrics of lung function. However, when subjecting the results to further analyses, I discovered that the post-race values remained well within the lower limits of normal, in nearly every instance. So, despite statistical significance reported in every study, the results weren't deemed to be *clinically* meaningful. The distinction is crucial because, from the data available, it appears that physicians need not be overly concerned with lung function following most marathon and ultra-marathon races, assuming the participant begins the race with robust baseline values, and they don't have any pre-existing respiratory disorders. Of course, there are always outliers and anomalies, and some individuals may develop respiratory abnormalities following arduous exercise, but it's not the norm for healthy people. This wouldn't be reflected in the findings of the original studies. Similarly, research into the ergogenic properties of a supplement might exhibit a statistically significant improvement in performance compared to a placebo, but it may translate to a negligible improvement in real-world outcomes. Be prudent, therefore, with respect to the *practical implications* of findings, and make an attempt to frame them in the broader context, should the data allow.

4.7. Other Resources

For informational purposes, the published, peer-reviewed scientific papers are your best resources. They're screened for quality and robustness (albeit sometimes in a cursory manner), and because new findings are always being published, the literature remains up-to-date. By contrast, blogs and websites aren't subject to the same scrutiny (there are no restrictions to online content), while books have the tendency to date rather quickly. But if you're to use a resource other than peer-reviewed papers to research a subject or product, some guidance on how to do so effectively is offered below.

Internet. Many commercial websites and blogs are independently run, meaning they have no formal processes of quality-control, and social media functions by generating content based on your previous viewing habits and those of your like-minded peers; accordingly, an echo-chamber is created in which you're rarely exposed to a full complement of information. Some websites exhibit greater credibility than others. A web address ending *.edu* belongs to a United States-affiliated higher education institution; that ending *.ac.uk* is the UK equivalent; websites ending *.gov* are government agencies. All are subject to stricter publishing controls. However, websites ending *.com* are commercial entities, and should be treated vigilantly.

Books, newspapers, and magazines. The correctness of these entities can also range from objectively true to dismally false. You must strive, therefore, to judge the information independently, but also consider the author on their credentials, experience, and reputation. They should be educated in the area about which they're writing, preferably have a degree or equivalent qualification, and belong to a credible organization. Objectively investigate the content of their writing and don't assume it's correct just because it reinforces your pre-existing notions. Isolate the words from the author; i.e., irrespective of the author, consider if the content is objectively true, and whether the legitimacy of the content would be altered if it'd been written by someone with whom you'd usually disagree. This is the true test of an objective, unbiased perspective.

Citations. Irrespective of the source – a journal article, a magazine, or a blog post – note if the author integrates one or more references into their commentary. Although blog posts themselves are not permissible as citations in laboratory reports, they're occasionally written by legitimate professionals; they're sometimes well-researched, coherently structured, and provide external links to journal articles and other reputable resources. You can isolate these additional papers to aid in your exploration of a subject. Wikipedia is a fine example of a site which, despite their stricter controls on content editing, isn't considered a reliable authority for the purposes of a scientific report. Nevertheless, the articles contained therein are often heavily referenced, and there's nothing prohibiting your independent review of those references; often, these citations are peer-reviewed documents, or reports from official governing bodies.

4.8. Ask...

To conclude the chapter, I'd like to comment on the importance of *asking*. For the non-expert, comprehending the sheer amount of data and information on a subject can be troublesome, to say the least. Reading a single paper is an elementary task if one has time, but examining a number of papers and balancing their arguments (sometimes opposing), while simultaneously deciphering the methodological nuances of each study, may be an unreasonable expectation. The first *ask*, therefore, pertains to asking an expert. This could be someone you know who's a specialist in the area, the colleague of a friend, a legitimate professional with whom you made contact through an online forum, or the author of a book you read (again, assuming they're appropriately credentialed); just ensure it's someone you trust. It's the expert's job to have a robust and comprehensive understanding of the subject, particularly if they're a researcher contributing to knowledge in the field. Why spend hours, if not days and weeks, carefully thumbing through journal articles when there's an academic or experienced practitioner close-by who's already done the heavy-lifting, and who can give you a verbal overview of the area? They'll also appreciate the subtler, as well as the more controversial, arguments that might be lost on a novice. Strive as best you can to grasp the subject yourself, but acknowledge the appropriate time to defer to an expert.

The second *ask* is a request for you to ask the manufacturer. There's a constant stream of exercise-related products appearing in the market, all connected to a series of claims with which we're bombarded. Contact the manufacturer and obligate them to justify a statement, or enquire after the evidence they have to support their claims. You can write a letter, send an email, or catch them unawares with a call. If there's any evidence to support their premise, they'll have it readily accessible, and should be amenable to sending you free copies of the reports. Only by demanding more evidence from manufacturers might we compel them to change the way their products are marketed. Be polite and genuine; the clerk with whom you'll likely converse isn't a despicable, wealth-obsessed tycoon, and you're far more inclined to succeed by being affable. But by the same token, don't be easily dismissed.

We've spent the most part of this chapter overviewing the inadequacies that may become apparent when examining products, and the specifics of how to read and interpret scientific papers which, when assessing product efficacy, serve as the evidence gold-standard. The higher-quality papers are those that control for external variables; in doing so, they minimize the inherent biases that can contaminate and distort the interpretation of data. Because humans are inherently biased creatures, such studies are critical in obtaining a clear and lucid understanding of *real effects*. We're afflicted with myriad means of self-deception, as well as an innate propensity to construct narratives which conform to our pre-existing notions of reality; i.e., reality is subjective. In the next chapter, we explore this concept more deeply. In addition to examining studies on the placebo effect in health and performance, we'll also explore how to minimize bias in order to better distinguish real effects from the fake.

5

PLACEBO PRODUCTS AND THE POWER OF PERCEPTION

per·cep·tion.

The faculty of perceiving, or apprehending by means of the senses or of the mind; cognition.

5.1. Intuition versus Intellect

I'll spare you a long and intricate retelling of Plato's *Allegory of The Cave*, but his narrative of three men forced to stare at a cave wall, having seen nothing during their lives except shadows cast on it by people on the outside, is an engaging hypothesis on human perception. In *The Allegory*, he claimed that human sensory knowledge is nothing more than subjective, opinions that are heavily influenced by our experiences. Our belief systems, in turn, influence the decisions we make but, according to Plato, real knowledge can only be attained through philosophical reasoning. As a prelude to the coming chapters, in which we confront head-on the health and performance claims of myriad products from nutrition to alternative therapy, presently we'll explore the opposing processes of intuition and intellect, and the bias that leads people to perceive positive effects of a sports product when no such effects exist. We also discuss why mitigating bias (as far as we can) is a prerequisite to formulating objective opinions on training programs and practices, nutrition, supplements, and alternative therapies.

Intuition and intellect are two profoundly different ways of *knowing* a thing, but where intuition is akin to instinct and dependent on your gut-feelings, intellect requires data, evidence, and careful thought. It's intuition that's more readily misguided by bias. In his much-lauded essay *Mysticism and Logic*, philosopher Bertrand Russell suggests: *...insight [intuition], untested and unsupported, is an insufficient guarantee of truth...* He further discusses an opposition between instinct and reason that I'd like to explore here, briefly, because it's become increasingly difficult to justify using intuition alone to form an accurate and holistic worldview. Instinct leads us to new and revelatory insights about the world, but such insights might be fallible and frequently erroneous. Before the scientific revolution (which began around the mid-1500s), philosophers had scarcely more than their intuition and instinct on which to test their understanding and base their decisions about the world. At the time of the ancient Greeks, the most plausible explanation for lightning bolts shooting from the sky during a thunder storm was the wrath of Zeus. Similarly, in the West in the late 1800s, bystanders wouldn't question the notion of a powerful tonic triggering the miraculous walk of a disabled man. But our explanations have since become more sophisticated, underpinned by our developed understanding of the physical sciences. Times have changed, and supernatural or otherwise mistaken explanations for established natural phenomena are no longer justifiable.

In the modern world, we have knowledge and understanding of the nature of existence that was beyond the reach of our ancestors. We can conclude with relative confidence if a particular claim is a close description of the truth. No longer must we rely solely on our intuition or instinct to understand the nature of events; we have the means to know for sure. Instinct is important, but it must be tempered by reason, directed by the powerful force of intellect. Collectively, this means we must be cautious when concluding that we *feel* a benefit when using a new product; such a position is based on perception and intuition alone. It may be convincing to drink a Snake Oil tonic and perceive yourself to be stronger in training, or to don some new running shoes and feel faster, or undergo a session of cupping therapy and perceive the tenseness in your muscles to have dissipated. But we have more at our disposal than blind and untamed intuition and it's crucial we learn to integrate these additional tools. It's even less acceptable to be so unyielding when there's objective evidence contrary to your perception. Many exercisers fall for the fallacious *it worked for* me defense, and it can be shaky ground to challenge the validity of someone's personal experience. Indeed, you risk a defensive response which can prematurely end any meaningful discourse. But it's a crucial step in objective enquiry. It does raise a difficult question; to what extent can we disregard such personal revelation when there's direct evidence to contest it? Can we really dismiss the subjective testimonies of friends and colleagues, risking personal and professional relationships? All we can really conclude is that if we use a product that makes untestable claims on effectiveness, then we must be comfortable to do so having rejected both reason and evidence. As Russell put it:

> Where instinct and reason do sometimes conflict is in regard to single beliefs, held instinctively, and held with such determination that no degree of inconsistency with other beliefs leads to their abandonment. Instinct, like all human faculties, is liable to error.

With respect to the vast array of natural phenomena, there's still so much left to discover, and it's likely that mankind will forever be searching for a true understanding of the nature of existence. But with respect to concerns like training programs, supplements, placebos, sports products, shoes, socks, and salt tablets, we have the ways and means to make definitive, educated decisions; i.e., we can apply intellect instead of intuition. And while only the robots and aliens in popular science-fiction – Asimov's R. Daneel Olivaw, or perhaps Mr Spok – are capable of the cold, inhuman logic required to dispassionately interpret the world and make purely objective, evidence-based decisions, the rest of us must accept that we are subject to emotive contaminations of thought. In such matters of establishing *truth* – as in delineating the factual basis for an argument or a claim – all we can do is our best to remove and mitigate bias at every step.

5.2. Bias

Imagine strolling down a busy high-street in a town near where you live. There's a bustle to the sidewalk, and every few seconds, a dozen shoppers rush past your flanks. Their shopping bags clatter as each person jostles for position in a vain attempt to claim a solitary space on the street. Noisy cars and trucks speed past on the road, spewing exhaust fumes into the air, and buses come and go dispensing and receiving passengers; while cyclists dart in between the vehicles, some have their backs laden with parcels destined for shopkeepers and businesses. The scenario is superficially chaotic, but the frenetic pace belies a structure and organization, at once order and disorder. Every stride has a purpose, every vehicle a destination, every sight, sound, and smell has both origin and demise. The sensory assault can be overwhelming for those people unacclimated to the city, while others thrive in the symphony of chaos. As the world speeds past, do you notice the registration plate of every car? Do you register the facial expressions of each passer-by? Or the size and shape of the parcel the postman was carrying, or the color of the carbon forks that sped past you a few moments ago? Do you notice each and every billboard advert, or the detailed constitution of storefronts as you glance through the windows? By contrast, do you think you'd more readily notice an attractive member of the opposite sex as they breeze past you, or if someone a few paces ahead dropped a wad of billfolds on the ground? What about two men engaged in an aggressive confrontation down the street; their voices raised in heated argument, their fists shaking at one another as a small crowd of onlookers pause from the daily grind and gather around the affair? Do these latter scenarios blend into the cacophony and the unrest to be concealed, or do they present themselves as ostentatious against the white noise of the city?

Each second of each day, your brain is filtering data, acknowledging and categorizing information it deems useful or important, and ignoring that which it doesn't (i.e., that which seemingly has no survival advantage). Given the vast amount of information with which we're bombarded daily, it's not possible to assimilate everything we see, hear, smell, and touch, and so filtering the detritus is a critical process to preclude overloading the higher cortical centers of the brain with irrelevant stimuli. Indeed, safely navigating the environment is predicated on knowing what information is *irrelevant* and what is *important* to our survival. This is a product of millions-of-years of evolutionary pressure and our brains have become adept at sifting through the noise and constructing a narrative that we can comprehend; in psychological research, this filtering mechanism is called *sensory gating*. It's an essential faculty without which we'd be unable to live.

Our filtering system is influenced heavily by our life experiences, predetermined biases, and beliefs. There is no reality other than that which your brain interprets; what's real is what you perceive. This is a fundamental obstacle to interpreting evidence and determining if the positive effects of an intervention are real or imagined. But a basic awareness of your inherent biases is the first step in learning to mitigate them.

A cognitive bias is a mistake in reasoning (or other cognitive process) that arises because we're reluctant to relinquish our personal beliefs or preferences, even when presented with sound contradictory arguments; marketers rely on such cognitive biases to influence your buying decisions. Biases are the reasons we cannot rely solely on our intuition when navigating the world. There are even manuals and online resources dedicated to assisting marketing companies exploit our biases to sell product. Accordingly, making objective, educated decisions is predicated on a clear understanding of the myriad ways that biases manifest in our daily lives. Here is an abridged overview of the main sources of bias with which you should be familiar:

Confirmation bias. This is the tendency for an individual to favor information that confirms a pre-existing belief, while simultaneously ignoring that which contradicts it. As such, we're more likely to buy a product or practice if we already *believe* it'll exert a positive influence on health or performance. Confirmation bias is evident in those people who only hear what they want to hear, and we've all likely been guilty of such an infraction. Hypnotherapy, for example, is a common means of assisting weight-loss or managing a physical ailment. The effectiveness of the treatment depends wholly on the client's pre-existing *belief* in a positive effect, and it'd be altogether ineffective on somebody who considered hypnotherapy to be squandered time. Often, no amount of evidence or logic will be sufficient to overcome an engrained belief, and what's more, that belief will be aggressively defended even when challenged by convincing evidence to the contrary. Confirmation bias is intertwined with ego – not the Freudian manifestation – but an overt defense of self-efficacy; one will protect it at all costs. We discuss ego, and the importance of humility in learning, in the final chapter. But when an individual is persuaded that a given intervention is the sole arbiter of their 10-km personal best, it may be beyond your powers of persuasion to convince them otherwise. As writer Jonathan Swift attests: *it is useless to attempt to reason a man out of a thing he was never reasoned into.*

Loss aversion. There's research to suggest that buyers fear losing something more than they regard gaining something of similar monetary value. A common strategy in marketing is an offer of the free-trial, and it's based on the premise that a *fear of loss* will make it difficult to give up when the trial expires. Each new calendar year, many gyms and health clubs offer free-trials because it's easier to convert long-term subscriptions this way. Rather than integrating this contrivance into your life (which leads to an inevitable sense of loss), the most effective way to overcome this bias is to adopt the attitude of *easy come, easy* go; consider the trial a temporary bonus that you'll exploit for a month before discarding (or simply convert to a full-time membership and be sure to use it). Feeling that you're about to lose out on a valuable, time-limited offer is another manifestation. A dietary supplement on sale, from $50 to $30, provokes an urgency to invest in something that's otherwise surplus to your needs because one is fearful of losing out, despite having scarcely paid $30 for the same product in any other instance. The offer creates an illusion of loss.

Congruence bias. We're all prone to jumping to conclusions. Congruence bias is to readily settle on the solution to a problem (even if it's incorrect) without acknowledging that there may be alternatives. Assume you have a problematic knee which becomes intermittently inflamed following exercise, and the pain abates when you place some insoles into your running shoes. As far as you're concerned, the problem is solved, although occasionally the pain returns a little less fervently. It may be that the nature of the injury is that it subsides periodically, or the insoles may have offered a temporary placebo effect, or you've reduced your training frequency, thereby allowing the knee to recover. The congruence bias manifests when the insoles appear to have corrected the issue, and you terminate your search for the true cause. A medical doctor at my running club recently exhibited this exact case. A characteristic of congruence bias is that people tend to construct tests that'll prove their hypothesis instead of tests that might falsify it. Indeed, it's more likely that one will persist with the insoles instead of trying to disprove their efficacy by removing them, or trying an alternative.

Exposure effect. This bias pertains to the visibility of a brand which, in turn, results in the inference that it's a superior product. Brands like Nike and Adidas are ubiquitous with sport; they sponsor athletes, shirts, and sporting events, and the exposure of their branding causes us to conflate their product with quality. Big-name brands sell a great deal more products than smaller, independent ones, even if the latter were to produce products of superior quality. This bias is also characterized by brand loyalty; we're more likely to buy products from brands that we know because they're safe and we know what to expect. The brand, therefore, occupies prime real-estate in our minds.

The Sunk Cost Bias. The more time or money one invests in a product, the more they'll be convinced that it was a good investment. It also becomes more difficult to abandon, even when presented with unfavorable evidence. The absolute financial cost of a product appears to play a major role in the perceived benefit. The phenomenon was nicely characterized in a 2008 study[17] which won the coveted Ig Nobel prize, awarded for *achievements that make people laugh, and then think*. Participants with chronic pain were randomized into two groups; one group was informed that they'd be treated with a high-priced medication ($2.50/pill), and the other that their treatment would be a low-priced alternative ($0.10/pill); in fact, both groups received inactive placebos. Pain relief was rated consistently higher by the group receiving the high-priced placebo, because of an expectation of more potent outcomes. In many instances, you'll continue to invest time and/or money in a product to justify the initial decision. If you buy an expensive pair of running shoes that bestow you with blisters and/or injuries, you might continue with their use long after you'd have tossed them away had you received them as a gift. You might pay eight months of a 12-month gym membership you've never used, and rather than simply cutting your losses and losing the remaining four months, it's psychologically more amenable to continue paying on the premise that you may eventually go. It's hard to give up on something in which you've invested a great deal of time. But the amount of time you've invested doesn't speak to the effectiveness, or factual basis, of the intervention.

There are exceptions to the manifestation of this investment bias. An analysis of 134,000 online reviews for running shoes from 24 brands (at www.runrepeat.com) found no association between shoe cost and overall rating[18]. In fact, although no formal statistics were performed, there appears to be a modest negative trend showing slightly lower scores for more expensive shoes (see Figure 5.1). I'm hesitant to make firm inferences from the data since the

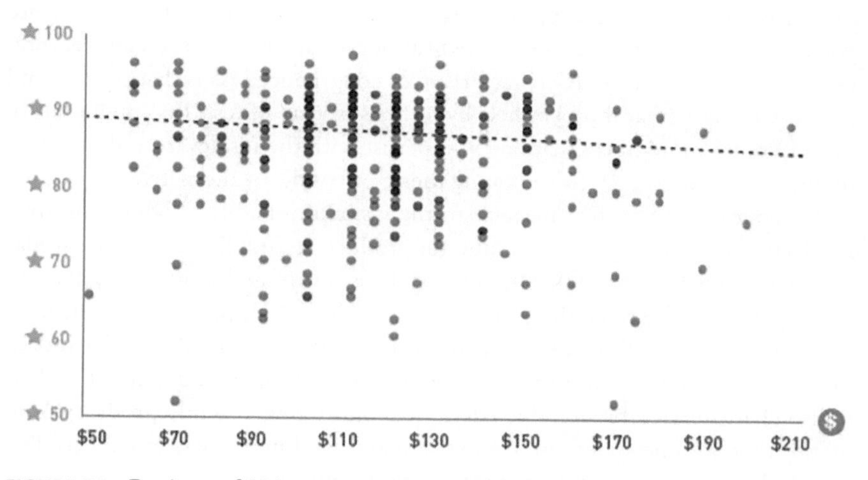

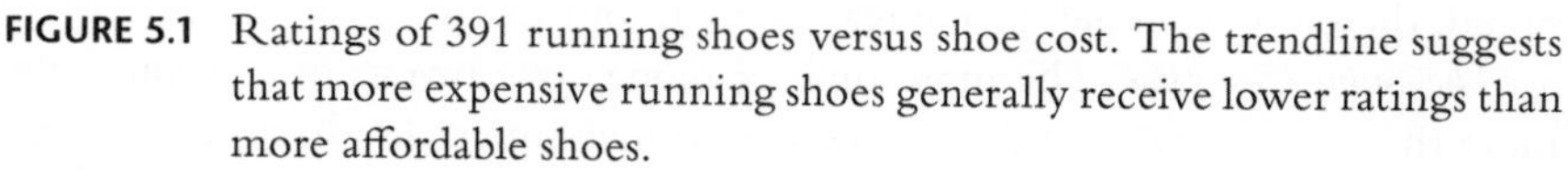

FIGURE 5.1 Ratings of 391 running shoes versus shoe cost. The trendline suggests that more expensive running shoes generally receive lower ratings than more affordable shoes.

Source: From runrepeat.com, used with permission.

study wasn't subjected to peer-review. Moreover, the analysis may have been biased by those people who tend to leave online reviews for running shoes and, as such, the sample may have self-selected. Another confounder is that spending more money on trainers might lead to exaggerated expectations and, therefore, greater potential for disappointment if said shoe confers anything other than dramatic ergogenic outcomes. Nevertheless, the site is independent (i.e., not owned by a manufacturer) and, at the very least, there was no positive association between shoe cost and star rating. Accordingly, investment bias may influence perceptions both positively and negatively.

5.3. The Placebo Effect: A Historical Perspective

Broadly speaking, there's substantial discrepancy between the commercial hype surrounding most health and fitness products and the evidence supporting their use. When a product has little supporting evidence-for-efficacy, it means that they've rarely been shown to exert any meaningful physiological effect on the body. But when a product improves performance via a positive psychological effect – attributable to an individual's *belief* in its effectiveness – the phenomenon is known as the placebo effect; it's ubiquitous with health and performance outcomes. It may seem like a minor contrivance, but the placebo effect can strongly influence our perceptions, and is not to be underestimated. Historically, placebos were considered a means of elucidating the validity of a medicinal intervention, i.e., whether the effect of a new drug or other treatment was due to a direct effect on the patient's physiology, or instead a result of their imagination. For example, imagine a patient is given a pill that they're told will reduce pain and relax their mood but, instead, the pill is a harmless, inactive compound that'll have negligible effect on their physical function. Before the invention of ether anesthetics in the mid-1800s, such a pill may have been administered before major surgery as a means of tempering the patient's violent responses to pain. Such deceptions weren't just necessary; they were humane, for medicine in the middle-ages was primitive and ineffective by modern standards. There were few other means at the disposal of physicians to assist patients through their painful procedures. In medicine, making such a distinction between real and imagined outcomes (with absolute assurance) is crucial for contriving an effective treatment. Consider that many patients simply won't recover without a physiological intervention.

If you're in any doubt as to the potency of placebo to effect perceptions of health and/or sporting performance, you'll likely be convinced by the end of this section. In his 1942 paper entitled *Voodoo Death*[19], Walter Cannon of the Harvard Medical School wrote: *When subjected to spells or sorcery or the use of 'black magic' men may be brought to death.* Cannon cited a series of examples from Africa, South America, Australia, and New Zealand in which individuals ranging from primitive tribesman to military solders suffered fatal consequences from believing they'd been damned by a voodoo curse. In many instances, Cannon recites *no nourishment or medicines that were given to him had the slightest effect either to check the mischief or to improve his condition in any way.* Although they hadn't the means to test the hypothesis, Cannon postulated that the fatal outcomes encumbering the patients were due to shock and fear: *shock and fear produce heightened activity of the sympathetico-adrenal system which lowers visceral blood volume and may produce deterioration of the vital organs.* Notwithstanding the wonderful prose of the time (*check the mischief* a somewhat understated way to describe a man dying of a purported voodoo curse), these testimonies offered an early insight into the mind's immensely powerful capabilities. If expectations – through the manipulation of an expected outcome, prior anticipation, and confirmation bias – are indeed capable of moving a man to death, it seems reasonable to suppose that outcomes of lesser magnitude, like improved exercise performance, may be easily achievable via the same mechanisms.

One of the earliest reports of intervention-based placebo effects (although it wasn't labeled as such until over 100 years later) was described in John Haygarth's 1800 book *Of the imagination, as a cause and as a cure of disorders of the body* (see Figure 5.2). In the publication, Haygarth recounts a clinical trial he designed to test the effectiveness of the so-called metallic *tractors* that had been developed by Dr Elisha Perkins. These *tractors* were small rods composed of various metals (the specifics of which were secret) that were capable of drawing disease from the body. They were sold to treat a wide range of disorders. Haygarth proposed the following:

> The Tractors have obtained such a high reputation at Bath, even amongst persons of rank and understanding, as to require the particular attention of physicians. Let their merit be impartially investigated, in order to support their fame, if it be well-founded, or to correct the public opinion, if merely formed upon delusion... Prepare a pair of false, exactly to resemble the true Tractors. Let the secret be kept inviolable, not only from the patient but also from any other person. Let the efficacy of both be impartially tried and the reports of the effects produced by the true and false Tractors be fully given in the words of the patients.

The *false tractors* described by Haygarth comprised bone, slate pencil, and painted tobacco pipes, and proved to be equally as effective as the original metal rods. A proven sham, Perkins' tractors were consequently dismissed by physicians of the age; Haygarth published an account of his findings in the abovementioned book. The world needs more John Haygarths! It wasn't until the 1900s that the term *placebo* was incorporated more broadly into the scientific literature. The placebo effect in medicine, as a potent pain reliever, has been discussed comprehensively elsewhere, but here's a typical example. In a study published in the journal *Psychological Science* in 1996[20], 56 University students were provided with a liquid gel which they were informed was a topical anesthetic; the gel

THE FAMILY REMEDY;

OR,

Perkins's Patent Metallic Tractors,

FOR THE RELIEF OF TOPICAL DISEASES OF THE HUMAN BODY;

AND OF HORSES.

Juſt Publiſhed, Price ONE SHILLING;

THE EFFICACY OF PERKINS'S PATENT METALLIC TRACTORS,

Exemplified by 250 Caſes on the HUMAN BODY, and on HORSES; in which a late Attempt of Dr. HAYGARTH, of Bath, to detract from the Merits of the Practice, by aſſerting its Action on the Imagination, is fully confuted,

BY BENJAMIN DOUGLAS PERKINS, A. M.

OF LEICESTER SQUARE, SON OF DR. PERKINS, THE DISCOVERER.

Publiſhed by J. JOHNSON, St. Paul's Church Yard; WRIGHT, Piccadilly; DARTON and HARVEY, Gracechurch Street; OGILVY and SON, Holborn, London; CRUTWELL, Bath; BELL and BRADFUTE, Edinburgh; and W. GILBERT, Dublin.

THE Complaints moſt ſubject to the Influence of the TRACTORS, are ACUTE and CHRONIC RHEUMATISM, including LUMBAGO and SCIATICA; GOUT, SPRAINS, CONTUSIONS, BURNS, SCALDS, INFLAMMATIONS of the EYES; alſo of the SKIN, as ERYSIPELAS and TETTERS; Painful INFLAMMATORY TUMOURS, as BILES and WHITLOWS; violent SPASMODIC CONVULSIONS, as EPILEPTIC FITS, CRAMP, and LOCKED JAW; PLEURISY; STINGS and BITES of VENOMOUS INSECTS; FLUOR ALBUS; PAINS in the HEAD, FACE, EARS, BREAST, SIDE, BACK, and LIMBS. ——The Diſeaſes of Horſes, ſuch as correſpond with theſe.

The very rapid increaſe of the reputation of the Metallic Tractors, their unperiſhable nature, the ſimplicity and innocence of their operation, even in the hands of the moſt illiterate, their eminent tendency to diminiſh the number of Medical Fees in the numerous familes in this kingdom, where they are owned, added to the well known univerſal maxim of the Faculty to oppoſe every Patented Remedy, neceſſarily expoſe them to virulent attacks and artful miſrepreſentations. As theſe can be accompliſhed through no medium more effectually than in Reviews, and ſimilar Journals, it may eaſily be conceived that ſuch a maſked mode would be eagerly embraced, and that every argument and explanation unfavourable to the Practice, which great ingenuity and learning could ſuggeſt, would aſſuredly be employed. It becomes therefore expedient, that a few unmutilated Caſes, a few ſimple Facts, be here adduced, from among the many hundred of a ſimilar deſcription, which are publiſhed in the book above-mentioned, in order that men, in the habit of forming opinions from the exerciſe of their own reaſon, may have an opportunity of ſo doing, without referring to thoſe which are the reſult of prejudice and ſelf-intereſt.

Before entering upon theſe Caſes, however, it may be proper to obſerve, that it is by no means intended to ſuggeſt, that all Reviews have become the vehicle of ſuch attacks; ſeveral have faithfully reported the nature of the Caſes which have been publiſhed, and expreſſed their opinion on the merits of the Diſcovery; as the following Extracts will ſhew.——

"To moſt people in years, the infirm in every period of life, all thoſe who are confined by contracted poſtures to unhealthy ſituations and ſedentary habits, and eſpecially the weaker ſex, whoſe feeble frames are often debilitated and cruſhed by their condition in life; a remedy nearly immediate is here propoſed, and authenticated by characters who cannot be ſuſpected of ignorance or connivance with impoſture. Of all medical diſcoveries urged on public attention from motives palpably lucrative, we are naturally and properly ſuſpicious; as of noſtrums and arbitrary preſcriptions there is no end. Theſe Tractors, however, have become, from the wonderful effects aſcribed to them, objects of juſt and general ſolicitude. And we are glad, for the honour of the Diſcovery, and more for the amelioration of our common nature, *which they amply promiſe*, to ſee them brought forward in ſo queſtionleſs a ſhape."---*London Review, for Nov.* 1799.

"The caſes which have occurred in England, are very numerous, and of reſpectable authority.—— Among theſe we have ſome of a different claſs from thoſe hitherto publiſhed, viz. upon Horſes. It would be invidious not to admit, that theſe reſults muſt effectually do away the idea, that the Tractors produce their effects through imagination. One gentleman obſerves, that he could perceive a ſwelling diminiſh, when uſing them upon his hunter, which met with ſevere contuſions by a fall. On the whole, we have no heſitation in acknowledging that the experiments adduced in this work, are of ſuch a nature, and ſo reſpectably atteſted, as to go very far towards removing the doubts we formerly expreſſed of the efficacy of this novel practice."—*Novelties of Literature*, (No. CCCXXIII.) *Oct.* 1799.

"Suſpicion never fails to attend every publication, which relates to an article ſold by the author. In the preſent work we ſee nothing which ſhould induce us to doubt the facts ſtated by Mr. Perkins. Theſe Metallic Tractors have been tried by many ſcientific characters in America, as well as in ſeveral parts of Europe. They are now ſold, under a Patent, in England, where they have alſo been ſucceſsfully uſed in many inſtances; and it would be at once cruel and unjuſt to ſuggeſt on ſlight grounds, and ſtill more ſo without any ground at all, ſuch reflections as might tend to leſſen the reward due to a meritorious invention." After a long quotation, they thus conclude. "Indeed of diſeaſes where relief has been obtained by the uſe of the Metallic Tractors, there are ſo many caſes in this work authenticated by well known reſpectable characters, that it muſt be the extreme of audacity and folly abſolutely to deny their efficacy."—*Weekly Review, Oct.* 1799.

FIGURE 5.2 Despite being exposed as a fraud by John Haygarth, public support for the metallic tractors remained steadfast, thanks in part to refutations like this from Benjamin Perkins, son of Elisha Perkins who had 'discovered' the tractors.

Source: Image courtesy of the U.S. National Library of Medicine.

rather comprised water, iodine, and oil of thyme to afford it a medicinal smell, but no active pain-relieving ingredient. The brownish concoction was applied liberally to the index finger of one hand, after which the index fingers of both hands were positioned in strain-gauges applying 2 kg of weight. Following the test, when asked to describe their experiences, the students unanimously sustained less pain in the placebo-treated finger. Worthy of note is that the finger selected for treatment (right or left hand) was randomized, and while half the cohort received the pain stimuli simultaneously on both fingers, the other half received stimuli to fingers in a sequential manor in alternating order. Studies such as this highlight the powerful influence of placebo in the context of suggestibility and outcome expectation.

For this reason, contemporary studies should test the effects of a drug or supplement by comparing outcomes to that evoked by a *placebo*. In practice, at least in cross-over studies, participants are provided the real drug/supplement on one occasion, and an inert placebo (perhaps a sugar pill or sodium chloride) on another, and the outcomes (the dependent variables, e.g., perceptions of pain, enzyme concentrations, exercise performance) are compared between trials. A treatment group that's *blinded* hasn't been poked in the eye with a sharp stick; it refers to a placebo with a similar superficial appearance and/or taste rendering it indistinguishable from the test product. Moreover, the nature of the product being received has been concealed from participants. With such a method, researchers can distinguish between real physiological effects and psychosomatic ones. In fact, the presence of a placebo group is one of the principal benchmarks that determines high-quality, well-controlled experimental studies. Strong studies may be *double-blinded*, denoting that both the researchers and participants are unaware of which product has been administered and on which occasion; this serves to eliminate bias on both sides, making the study more robust.

Before overviewing the widespread use of placebos in sport, it's important to understand the basic mechanism of the phenomenon. When we consider the context in which placebos are observed in the health and fitness industry, there are important questions that arise: how does the placebo effect deceive us into believing we are experiencing a real physical phenomenon? Why does the expectation of a positive result influence our perception? What actually happens in the brain and body when we're subject to (or fooled by) a placebo, and how does this improve our perceptions of health and/or performance?

5.4. How Do Placebos Work?

What follows is a brief summary of a deeply complex series of interrelated mechanisms; as such, it's not an exhaustive review. Nevertheless, a superficial understanding of the placebo effect and how it alters decision-making will serve us well in the next section during which we look at sporting examples. The two main concepts that underpin the placebo effect are suggestions/

expectations, and a type of learning called Pavlovian Conditioning. If you're a psychology graduate, you'll likely already have a strong comprehension of the phenomena, and if not, it'll probably resonate with your high-school education. Such learning was popularized in the late 1800s in experiments by Russian physiologist Ivan Pavlov. In the course of his studies, Pavlov discovered, rather intuitively you might say, that his dogs began salivating at the sight of food brought to them by his assistant. Pavlov determined that salivation was an unconditioned response to the unconditioned stimulus of food; i.e., the animals had no need to *learn* to salivate at the sight of food, as it was a genetically-predetermined response inherent in most mammals (humans included). With time, the dogs would salivate not at the sight of food, but at the sound of the assistant's approaching footsteps. The breakthrough emerged when Pavlov conditioned his dogs to salivate on hearing the sound of a metronome which he activated immediately prior to feeding time. Eventually, the sound of the metronome alone (in the absence of food) was a sufficient stimulus to trigger salivation. The metronome had become a conditioned stimulus evoking a conditioned response; i.e., the dogs had *learned* a new behavior, that is, associating the sound of the metronome with food. These are the basic principles of classical conditioning, and they underpin the earliest systematic studies of learning. The precept is that such learning is based upon associating an unconditioned stimulus (e.g., food or a potent drug) that already causes a particular response (e.g., salivation or improved performance), with a new conditioned stimulus (e.g., a metronome or a placebo), so that the new stimulus evokes the equivalent response; consequently, your placebo product, which has no direct demonstrable effect on physiological function, results in a faster time-trial, greater jump-height, and improved energy levels.

A placebo intervention might trigger subtle associations of places, persons, procedures, or other products previously evoking the same outcome. But the Pavlovian conditioning and suggestion response is a single piece of the puzzle. With respect to pain management, placebos not only influence the pre-stimulus expectation of pain, but also modify pain perceptions, and the ratings of post-stimulus pain. Moreover, the large amount of literature on the topic suggests that multiple brain regions (including the anterior cingulate cortex, anterior insula, and the prefrontal cortex) play an important role[21].

Indeed, a 2002 study exposed nine healthy individuals to a non-harmful 'painful' experience[22]. They received a real analgesic (pain-relieving) injection, an inert placebo injection, or a no treatment control. Although the analgesic injection provided the most effective pain relief, subsequent brain scans (PET) determined that the rostral anterior cingulate cortex (associated with a variety of autonomic functions including pleasure and reward) was active following both the treatment and placebo injections. Accordingly, despite no demonstrable *cellular* mechanisms, placebo mediates symptoms of pain instead by influencing brain activity. It's currently unclear if the same brain regions active during placebo pain management are the same regions that contribute to exercise performance, but it's likely that areas like the striatum – responsible for reward – are pertinent.

So, placebos combine strong expectation biases with changes in brain function to exert very powerful effects. In the present context, such outcomes must be kept in perspective relative to other interventions. Indeed, where a given supplement (e.g., Beta-alanine, caffeine, or glucose) has been documented to exert a real ergogenic benefit (i.e., non-placebo), the magnitude is reported to be 1%–3%[23,24,25]. This is similar in magnitude to those effects observed with positive *beliefs* triggered by placebos. Accordingly, the premise on which the sports supplement industry is founded may need to be re-evaluated, given that, in some cases, a similar performance benefit may be obtained with the use of both real and fake supplements. These speculations must be carefully considered, and studies contrived to directly compare supplement and belief effects.

5.5. Placebo Effects in Sport

The notion of placebo is widely endorsed by both athletes and coaches. A 2007 survey of 30 athletes revealed that 97% were of the opinion that the placebo effect could positively influence their performance, while 73% felt that placebo caused them to make a deliberate change in their competitive strategy, altered their belief in the expected outcomes, or tapped into their belief in marketing or faith in a third party[26]. Extending these findings some years later, a 2015 survey of 79 elite athletes from a range of sports determined that 82% of respondents thought that placebos could affect their sports performance[27]. Intuitively, athletes were significantly more likely to endorse placebos in sport if they'd used them previously and had a positive experience, reinforcing the notion that a conditioning effect is congruent with a deep-rooted expectation of positive outcomes. This link forged by the athlete (between the placebo and the expected outcome) then activates various neurobiological mechanisms which lead to the beneficial outcomes. Importantly, the efficacy of placebo hinges on the successful delivery of a deception, and be mindful that athletes rarely work in isolation in competitive sport. Indeed, the buck usually stops with the head coach and, accordingly, their attitudes toward placebos have also been studied.

The perceptions of 96 coaches from regional to international level were reported in a 2016 study[28]. Approximately 44% disclosed that they'd previously tried to influence their athletes' performance by administering a placebo. Interestingly, the practice was more common at higher levels of competition (Regional = 30%, National = 35%, and International = 60%). Crucially for this chapter, an overwhelming majority (93%) of coaches utilizing placebos reported some ergogenic outcomes in their athletes.

Placebo effects on sports performance have been well-assessed using innovative study designs. Here are some examples from the sports supplement industry, using nutritional aids that are widely regarded to positively influence performance, and for which the evidence-for-efficacy is fairly strong. The first study recruited 43 competitive cyclists to complete two 40-km time-trials in a laboratory[29]. On the first occasion, the cyclists were given only water so that a *baseline* performance could be determined. On the second, they were given either a drink containing an 8% glucose polymer (which is similar in concentration to most commercially available sports drinks), or they were given an identical-tasting glucose-free placebo. In the latter trial, the groups were further divided into those informed they'd been given glucose, those informed they'd been given the placebo, or told nothing at all. With such a design, the study was able to determine the magnitude to which expectation influenced cycling performance. As an abridged summary, the carbohydrate (glucose) supplement was impotent to effect performance, as might be expected in a time-trial of such short duration. By contrast, the most improved performance relative to baseline (+4%) was exhibited by the group who were given flavored water but informed it was glucose. This study offers a lucid insight into the very powerful effects of placebo on sporting performance.

A second example from the literature explored the notion of caffeine as an ergogenic aid, its methods altogether more cunning[30]. Six well-trained cyclists performed a 10-km time-trial in the lab, on three separate occasions. On each occasion, they were informed that they'd be given either a placebo, a moderate caffeine dose, or a strong caffeine dose; in reality, however, none of the supplements were caffeinated. What emerged was a dose-dependent response with respect to perceived caffeine intake; i.e., power output was augmented by 1.3% when cyclists believed they'd taken a moderate dose, and by 3.1% when they believed they'd consumed the strong caffeine supplement. All subjects reported caffeine-related symptoms. Power fell by 1.4% when knowledge of placebo became apparent. An enhancement of 3.1% over the course of a 10-km time-trial is not a trivial one.

The expectation that comes from active ingestion of a perceived sports supplement isn't the only mechanism by which placebo can manifest; the perceptions surrounding altitude interventions have also been tested. Altitude interventions have gained a great deal of popularity in the last few decades due to a belief that living and/or training at high altitudes (>2,500 m) might stimulate adaptations in the body that facilitate endurance performance. A study in the *European Journal of Applied Physiology*[31] tested performance in highly trained cyclists before and after they'd spent 26 nights sleeping at simulated altitude (a five-bedroom facility in which the oxygen content had been artificially reduced via nitrogen dilution). The novelty of this particular study was that after 14 days of exposure, half the group provided weekly blood donations via phlebotomy which negated any further adaptations in hemoglobin mass; both halves of the cohort were blinded as to the volume of blood withdrawn. By the trial's conclusion, there'd been no alterations in hemoglobin mass in those subjected to weekly phlebotomy, whereas the other group exhibited significant increases from baseline. Despite physiological adaptations only in the latter group, cycling performance in a short-duration, high-intensity trial improved in *both*, with no significant differences between them. There are further details to this innovative study that warrant exploration in your own time, but the data suggest that the benefits of simulated altitude may be attributed, at least in part, to the expectation of performance enhancement that accompanies the intervention. Indeed, any intervention that's congruent with a confident or forceful claim to efficacy has the potential to exert non-specific (placebo) effects if the consumer buys into the premise.

5.6. The Price of Placebo

To conclude this chapter, I'd like you to consider more deeply the implications of placebo, and whether its powerful effects can truly be harnessed without repercussions. The early history of placebo revealed how credentialed professionals harnessed lies to achieve desired outcomes[32]. Administering a placebo to an athlete, exerciser, or patient requires their belief that the treatment is effective and their expectation of a physiological outcome. This is a dangerous precedent, because *any intervention* can work in the context of placebo. Intuitively, either we accept all products that claim to work in this context (rendering us susceptible to a further cascade of products without proven effects), or we accept none of them. But there may be specific consequences to the use of placebos in both the sporting and clinical settings. From a sports perspective, for example, there's often a close working relationship between practitioners and athletes. In both Olympic and professional programs, sports physiology and medicine for example, is fully integrated into the scientific support system; there's no two-week delay in obtaining a consultation with a doctor, no fees for the athlete, and support from medics can be accessed as easily as that from physiologists, nutritionists, psychologists, strength and conditioning coaches, and physiotherapists. Performance sport is a particularly high-pressure environment, with both athletes and coaches generally harboring a *win at all costs* mentality in which minute gains are additive, and accumulate with time in order to distinguish success and failure. There are positives to trying novel, trendy interventions; not only might we chance an expedited recovery from injury or overtraining, but the physician appears to be doing all in their power to facilitate a return to competition. But practitioners may feel pressure from athletes, head coach, or performance director to access contemporary *cutting-edge* treatments, even when such interventions lack convincing evidence to support use. The practitioner, therefore, must consider the ethical basis of prescribing a treatment that they know will only work in the context of placebo. On the one hand, many-a-sporting competition is won or lost in the athlete's mind, in which case a placebo intervention can provide a crucial psychological edge. On the other, these interventions can be costly, time-consuming, may provoke unwanted side-effects, and can detract from more deserving ones.

Nevertheless, if the athlete *believes* with sufficient conviction in the potency of an intervention – e.g., a pill, some taping, an alternative treatment – then there's greater propensity that it'll bestow an ergogenic effect. The strength of the outcome is dependent on the strength of the belief. Even if a product was proven after the fact to be inert, research has shown that 67% of elite athletes wouldn't scorn a placebo-mediated deception if it were effective[27]. Moreover, assuming a subsequent ergogenic effect, most of those surveyed said that they were amenable to being deceived by their coach into taking a placebo, and 10% (1/10) would take an *unknown* substance without questioning its effects or legal legitimacy. The latter statistic is particularly alarming, and draws attention to how some athletes consider the end to justify the means.

From a clinical standpoint, by managing patient expectations, the mind can be a powerful tool in overcoming pain and certain other dispositions. A minority of physicians offer the so-called *sham* treatments for minor ailments on the basis that the patient's *expectation* of a remedy will often suffice. Much of the success of homeopathic hospitals – one of which located in London, UK, initially benefitted from £10 million of government support before funding was withdrawn owing to the dire clinical evidence – is underpinned by the longer patient–doctor consultation times relative to that offered in traditional medical practices. The extra pastoral care afforded to patients in homeopathic clinics was crucial in influencing patient outcomes. Critical to this discussion is that when the attitudes and perceptions of over 6,000 individuals were surveyed, it was apparent that most wouldn't have felt deceived if a placebo treatment assisted their recovery from a hypothetical illness; moreover, they didn't think that the placebo would have been successful had they been fully informed[33]. Again, the data suggest that outcomes take precedence over procedural honesty.

There's no disputing the potency of its effects, and athletes, coaches, and patients concur that placebos are valid and efficacious, despite being attributable to psychosomatic phenomena. If so congenial and effective, then why question the widespread use of placebo? Well, a contrasting viewpoint pervades because of ambiguity regarding the long-term harm and ethical dilemmas that placebo use can evoke.

Many placebo-mediated therapies carry inherent risks, and scientists/physicians must consider their motives before recommending them. Proposing the widespread use of placebo is a slippery slope. On the one hand, lack of efficacy can have relatively benign consequences; if an athlete uses an inert supplement expecting ergogenic effects but receives none, then it's disappointing but hardly life-changing. On the other hand, the medical literature is littered with instances whereby minor ailments have progressed into serious health concerns because they weren't appropriately treated. A placebo might be effective if prescribed for mild pain relief, but for an acute bacterial infection, quite often nothing short of antibiotics will suffice, and no number of inert sugar pills can offer recourse. Treating an asthma patient with a sham intervention to mitigate their breathlessness may appear harmless, but it's axiomatic that treating an asthma attack with a placebo could be fatal. Indeed, patients frequently forgo modern science and medicine in favor of unproven and ineffective alternative remedies, with tragic consequences. In 2017, for example, a seven-year-old boy contracted an ear infection which would usually have been resolved with a course of antibiotics; however, the child's condition worsened to a coma, and he died from encephalitis (inflammation of the brain) following treatment with an inert homeopathic *medicine*. His parents, found guilty of his manslaughter, were punished with a meager three-month suspended sentence, while Massimiliano Mecozzi – the homeopath who had *underestimated the seriousness of the illness* – was scarcely reprimanded with a six-month suspension from medical practice. This example was one of numerous I could have cited, attributable to a belief in products with effects that are placebo-dependent. Cryotherapy is used for the purpose of treating sports injuries and reducing inflammation associated with arduous training and competing. However, while at one end of the spectrum proponents of cryotherapy claim that it expedites recovery, at the other they claim it as a cancer treatment. One can argue that sanctioning and condoning the use of placebo products tempt an avalanche of inappropriate claims, and it's a path that will lead to inevitable consequences.

On further examination, it's also apparent that the use of alternative treatments like acupuncture or cupping for health and sport (see *Chapter 9; Alternative Therapies*) means we have less time and money and other resources to devote to treatments that have clear mechanisms and supporting clinical evidence. The athlete's relationship with their practitioner (coach, sports scientist, or physician) is a critical element that underpins their belief in, and responses to, a placebo. These practitioners are, therefore, in a position of immense power, and with great power comes great responsibility. To advocate an intervention with no biological influence, fully cognizant that its effects are placebo, is to perpetrate a deliberate and profound deception. Moreover, placebo products in performance sport are scarcely provided in isolation; integration of a new supplement or treatment is discussed and reasoned, and agreed-upon by various members of the support team, including medics, psychologists, nutritionists/dieticians, physiologists, and coaches. As such, a placebo is predicated by widespread scientific dishonesty and intellectual cognitive dissonance. Many scientists, myself included, feel that intellectual and academic integrity is worth protecting, particularly in the modern commercialist culture characterized by an abundance of systems and institutions eager to see such integrity sacrificed for monetary gain.

By way of resolution, the answer may be to invest our resources in ergogenic aids (be they nutritional, physiological, mechanical) that are underpinned by both empiricism *and* powerful expectation/belief effects; scientists can, therefore, optimize performance while retaining their ethical standards[34]. An effective placebo is predicated on the athlete believing wholly in the ergogenic potential of the product, but the shaking of such blind faith causes the magical properties of the inert placebo to vanish. So, how do we decide that one claim is worthy of belief and another worthy of skepticism? The answer is evidence-based understanding, driven by the scientific process. Although some will retain the cogency of fake products in sport, the principles at stake supersede the sporting arena, pervading politics, the economy, and medicine. The pursuit of science is ultimately the pursuit of truth, and it may be that *pseudo*scientific practices in sport fundamentally hinder the progression of *true* athletic performance.

If, on some distant world, at some nondescript time in the hypothetical future, the sporting authorities rule that deceiving athletes is in violation of ethical governance and, hence, illegal, would the so-called placebo products become banned practices? And what would be the implications of banning a practice that cannot be quantified or detected? This isn't analogous to doping, as has been suggested by some[27] because the athlete has taken no action to *directly* influence their physiological make-up. Nevertheless, there are tangible outcomes to the use of placebos and, at the very least, such practices could be considered *misconduct* on behalf of the coach or practitioner, or the other entity selling the placebo, i.e., the manufacturer. Could we, one day, live in a world where a product with no physiological mechanism or quantifiable physical effect – sold on such a promise – could be illegal practice? Might there eventually be legislation to prevent manufacturers selling products with no efficacy beyond the expectation bias of placebo? It shouldn't be beyond our imagination to visualize such a time and place.

In this regard, consider that placebo products are entirely legal and there is no prohibition on their widespread use. As aforementioned, athletes learn to associate an intervention with an expected outcome. But consider the repercussions if a coach covertly replaced his athlete's caffeine tablets with an illegal performance-enhancing drug (a more potent stimulant, for example). Evoking a powerful assault on the sympathetic nervous system, the physiological response would be substantial. As the weeks passed, the athlete would forge strong associations between the drug and the response and, eventually, the potent outcome would be anticipated. Assuming both athlete and coach could avoid the anti-doping committee during this initial period, the stimulant could eventually be replaced with a harmless sugar pill, and the now conditioned athlete would reap many of the ergogenic benefits without the associated risk. Although speculative, such placebo pre-conditioning with banned substances may yield extremely powerful results, but would be chemically undetectable in conventional drugs tests.

While the argument wages on, it's apparent that the perspectives of the academic community are relatively less important, for sales are driven by consumers – the athletes, exercisers, and patients – voting with their credit cards; it's the consumer who dictates product popularity. We must engage scientists, physicians, patients, athletes, and coaches in ongoing dialogue to determine the best way to implement the use of placebo products, if at all. At the very least, there must be broader education on the potentially harmful implications of placebo, and a drive to equip consumers with the tools they need to inoculate themselves against potentially destructive *pseudo*science.

6

SPORTS NUTRITION

mar·ket·ing.

The total of activities involved in the transfer of goods from the seller to the buyer.

6.1. A Lucrative Industry

Good nutrition is crucial for health[35], weight-loss[36], and sports performance[37]. While most are familiar with the basic tenets of nutrition, the challenge of mastering it by implementing habitual healthy-eating can prove substantial. Accordingly, if tasked with designing a new dietary regimen, most would willingly defer responsibility to an expert dietician or nutritionist. This concession comes at a cost, however, because a chronic and widespread ignorance of nutrition serves as an ample invitation for self-confessed nutrition *experts* to hijack the market with fad diets, supplements, and contrary and sensationalist advice. Indeed, a vacuum created from a dearth of science and critical-thinking is rapidly filled by *pseudo*science and false claims. The world of sports nutrition and healthy-eating has become a confused and convoluted argument in which only the loudest voices are heard. Somewhat ironically, many have made their fortune in the industry by exploiting public confusion and harnessing the power of advice that's simple and easy-to-follow. While these New-Age gurus should be commended for making good nutrition once again accessible to the mainstream, much of their advice is anchored on fallacious appeals to nature and popularity. Moreover, there's no need for good practice to be commercialized and sold as if it were an innovative breakthrough in biohacking. Since when did a diet espousing low sugar necessitate its own website and hashtag? Since when did the consumption of fruit and veg need to be sold in a series of books and online blogs? It presently appears that the public only implement dietary tautologies like *avoid refined sugars* and *reduce alcohol intake* if the advice is packaged and sold as a *product*, alongside a marketing hook. Accordingly, this chapter explores the contentious world of sports nutrition. Therein, we'll focus on the claims stemming from proponents of fad diets like superfoods, juicing, and the infamous detox. We discuss the ethos of health food stores, and scrutinize organic foods which make health-based claims that are empirically testable. Finally, we'll confront the media's critical contribution to the perpetuation of bad science in the nutrition industry.

6.2. The Good, the Bad, and the Tasty

Regardless of how disciplined your diet, it'll be periodically interrupted with a vice like chocolate, alcohol, or the weekend takeaway. Such indulgences are a normal and some may say integral component of a healthy-eating regimen, but moderation is key if our diets are to enable, rather than impede, our health and/or performance. Many of the nutrition-based interventions targeted at exercisers (fad diets, supplements, detox kits) are predicated on the simple notion that we should be eating more *good* foods and eating fewer *bad* foods. However, I'd like to begin this chapter by challenging the validity of this concept, and assert that such binary categorization of foods is both meaningless and unhelpful. Understand that *unhealthy* foods don't exist; instead, there are only certain foods that can have long-term unhealthy *effects* on the body if disproportionately consumed. Consider a confectionary product like a donut. Few would argue the notion that this familiar blend of flour, sugar, and fat is *junk food*, and that we should abstain from regular consumption. But the donut itself is not inherently *unhealthy;* the label is a meaningless abstraction that changes with context. Now consider two cases of people at extremes of the physical activity spectrum: (1) a middle-aged, sedentary, overweight individual, whose physician has advocated a decrease in body fat to reduce the risk of poor cardiovascular health; (2) a high-performance athlete training to compete as a middle-distance runner at the Olympic Games. The first subject should generally abstain from donuts. These tasty treats are calorie-dense (predominantly sugar) and, on consumption, elicit a sharp spike in blood glucose concentration that triggers hormonal responses in the body that lead to the additional storage of subcutaneous fat. The subject's present condition is likely attributable to the excessive consumption of calories, combined with low activity-levels (although there are sometimes genetic causes). Their goal of losing body fat in the long term will be rendered a failure if their net calorie intake is not surpassed by their net calorie expenditure. In this context, the donut itself has negative outcomes if consumed too frequently. However, what if the subject has eaten meticulously all week, they've lost several pounds of body fat, and used the donut as a weekend sweet before persisting with their healthy-eating regimen; in such an instance, the sweet won't elicit a net harmful effect on the body and, by way of a reward system, might actually play an important role in long-term adherence to a healthy-eating program. The context has changed, and the validity of the donut as an inherently *bad* food is brought into question. The Olympic athlete perhaps trains 2–3 times per day, 6 days per week, requiring many thousands of daily calories to meet their energy demands and facilitate recovery from arduous training. Imagine the athlete has just completed a hard training session lasting around two hours. Unfortunately, they've left at home their recovery meal, and they're carrying just pocked-change. Moreover, they're about to board a train, thereby delaying their next meal by several hours. After training, their body is in a catabolic state, immune function may be suppressed, and the subject is hungry and irritable. If a donut purchased from the local shop were the only option, most would consider this a reasonable concession rather than going hungry. When the dichotomy is junk food or no food, the nuanced scenario dictates that the junk food has now become an important aid to recovery. Once again, context has brought into question the notion that donuts are inherently unhealthy. While acknowledging that the athlete could've opted for a banana instead of a donut, the precept is that foods cannot be classified as good or bad, healthy or unhealthy, only that they exert various effects on the body that are dictated by context and individual needs. We each have different demands, and establishing these at the commencement of a new regimen is key to success. Moreover, in the health and fitness industry, it's critical to question the principles that underpin our buying habits; this is a common thread running through the book.

6.3. Nutritionist or Dietician?

Few exercise professionals would argue that achieving health and/or performance outcomes depends heavily on adequately fueling your body with the right foods (note that I didn't use the term *good* foods). With such public demand for nutritional guidance, you're scarcely far from a self-professed *expert*, a fad diet, or nutritional supplement. But with so much advice around every corner, how do we succeed in interpreting the wealth of information (and misinformation) with which we're bombarded daily? The first step is to distinguish between those from whom we should, and should not, be seeking advice.

First acknowledge that not all *experts* are created equal. In general, consumers have a misplaced view of what constitutes an *expert*, and what might be considered an appropriate set of credentials. For example, if you needed an operation to repair a shoulder damaged through sport, and were to go under the knife at hospital, you'd expect your surgeon to be well-qualified for the task, and have sufficient experience. You might even want to know where they'd studied, worked, and their past patient track-record. If you were hiring a maths tutor for yourself (or a child), you'd expect the tutor to have substantial experience in education, and perhaps be affiliated with an appropriate governing body. If you were taking a flight on a business trip, you'd expect your pilot to meet the same standard of stringent criteria as your surgeon. Certainly, you wouldn't select a surgeon, teacher, or pilot based on their Twitter following, whether or not they had a verified Twitter account, or how many *likes* had accrued on their Facebook page. Yet, this is precisely how many people choose their nutritionists and exercise professionals. The first principle, therefore, is that Twitter followers are not credentials!

So, who can we trust to offer advice on healthy-eating? Most of us would seek out a nutritionist or a dietician, but recall that not all were created equal. To practice as a dietician, one must first have earned an undergraduate degree in Dietetics or related field, followed by advanced study that culminates in a master's (or doctorate) in the same subject. The trainee must then complete a period of supervised experience, usually in a hospital or clinic, and then achieve their accreditation to practice by aligning with national and international standards for professional legislation. The title *dietician* is protected by law in Canada, USA, South Africa, Australia, and the UK. By contrast, the title *sports/performance nutritionist* remains unrestricted; it's a designation available without any formal qualification, and the process is poorly regulated. Consequently, many personal trainers and gym instructors offer healthy-eating advice under a self-determined title.

Note that many nutritionists are registered with regulatory bodies, qualified with advanced degrees and/or doctorates in nutrition, and possess superior understanding of food science, often surpassing their dietician counterparts. But poor regulation of the title leads to a broad range of abilities, making it very difficult to distinguish those with genuine knowledge and experience from those with a gimmick and bravado. This also doesn't render incorrect the advice from a poorly credentialed practitioner; consumers must exercise caution and judge each practitioner individually.

6.4. Nutrition in the Media

Humans are obligated to exhibit an interest in their diet by virtue of the fact that eating is a crucial component of continued consciousness. A principal obstruction to the public understanding of nutrition (and science in general) is the nature by which it's often reported in the media. The value of reads, likes, retweets, and shares results in science often taking a backseat to sensationalist headlines; thus, we have a nation obsessed with superfoods, antioxidants, and detoxing, all of which are accompanied by grand health claims and poor evidence. The ease with which some journalists misrepresent the facts is attributed, at least in part, to widespread science illiteracy among consumers. On why he became a journalist, Christopher Hitchens exclaimed: *I became a journalist so that I wouldn't have to rely on the press to know what was going on.* In this vein, I present two news stories and one anecdote which illustrate the necessity of critical-thinking in the mainstream media. These instances are a general comment on the nature of science reporting.

6.4.1. Chocolate Accelerates Weight-Loss?

The first story is centered on a study published in the *International Archives of Medicine*[38]; it features in my Applied Physiology class because it's an excellent example of how deeper and more robust critical-thinking skills could protect you from the sensationalist reporting of scientific falsehoods. There were flaws in this study – from conception and design, to data-collection and analysis, and eventual publication – as well as in the subsequent reporting by the media. The essence of the report was that eating 42 g of dark chocolate daily, congruent with a low-carbohydrate diet, resulted in 10% faster weight-loss than a non-chocolate diet of equivalent energy. The headline *Slim by Chocolate* was proclaimed in 20 countries and 12 languages, including respected UK outlets like BBC News. In response to the media circus, and his newfound notoriety, the study lead-author Dr Johannes Bohannon – Research Director of the Institute of Diet and Health – said: *the best part is, you can buy chocolate everywhere!* This single (albeit preliminary) study was authentically performed; they tested real people adhering to real diets, and recorded real data. Nevertheless, unbeknownst to the media, Bohannon's study had been designed for the purpose of testing the critical-faculties and scientific integrity of the mainstream media, a test they assuredly failed.

Dr John Bohannon earned his title by studying the molecular biology of bacteria. The Institute of Diet and Health – where Bohannon serves as Research Director – is a façade anchored on a website designed by Bohannon and colleagues to authenticate his credentials. The *International Archives of Medicine* which featured the article is a pay-to-publish journal that charged 600€ to publish the paper online just two weeks after being received. The uncredentialed author, the fake research center, and the bogus journal were low-hanging fruit that should've been detected by any journalist prioritizing research validity above sensationalist headlines.

The sting was conceived by German television reporter Peter Onneken; he and his collaborator, Diana Löbl, were developing a documentary concerning the poor standards of science reporting in the diet industry. Bohannon was recruited to help demonstrate the ease with which bad science could be converted to front-page news. But the hoax – more sophisticated than at first appearance – comprised a real study with real participants and data, manipulated and molded to evoke the interesting find. Following the initial publication, Bohannon published a blog-post in which he described the statistical analysis and subsequent data-mining process[38]. Relative to the original story, the *big reveal* wasn't reported by the media with an equivalent gusto. Eventually, there was widespread coverage and the truth had its day, owing to diligent work curtesy of the vocal science community and science bloggers from around the world.

The study was of standard design; participants were randomly assigned to one of three experimental groups for a three-week period: (1) a low-carbohydrate diet, (2) a low-carbohydrate diet plus a daily 42 g chocolate bar, or (3) a control group in which there was no dietary intervention. After three weeks, it was the chocolate group who'd lost significantly more weight (10%) than the group following only the low-carb diet. The explanation for these observations comes by way of rudimentary inferential statistics. The study tested a huge number of variables – 18 in total – including: weight, body-mass index, waist-to-hip ratio, cholesterol, lipids, sodium, albumin, blood protein levels, sleep quality, urine ketone concentration, weight-loss, rate of weight-loss, and various perceptions of wellbeing. When all but the kitchen sink is thrown into the analysis, the chances of unearthing a statistically significant result, purely on the basis of random chance, are substantially enhanced. Any statistical test employed to make comparisons between groups comes with a small probability of producing a false positive called a type I error. Researchers tend to employ an arbitrary cut-off for an acceptable level of probability (the *p* value), usually 0.05 (5%). Accordingly, if the test reveals a significant result, one can be at least 95% confident that the effect is authentic. With such an alpha level, 100 comparisons among data sets would randomly yield five positive results (5%); 20 comparisons would be sufficient to almost guarantee a statistically significant result. Moreover, there was no attempt to amend the analysis to account for multiple comparisons, as is the norm. The study was, therefore, designed with the intention of delivering a significant result, and the statistical analysis was conducted in such a way that the researchers were able to uncover patterns in the data without first hypothesizing a mechanism. Making so many comparisons without appropriately adjusting the statistical analysis is called *data mining*, and it's very bad practice. In Bohannon's words:

> Think of the measurements as lottery tickets. Each one has a small chance of paying-off in the form of a "significant" result that we can spin a story around and sell to the media. The more tickets you buy, the more likely you are to win. We didn't know exactly what would pan out – the headline could have been that chocolate improves sleep or lowers blood pressure – but we knew our chances of getting at least one 'statistically significant' result were pretty good.

Notwithstanding the misleading statistics, there were further deficiencies in the publication process which complemented the deception. The online resource that published the paper didn't appear to implement a robust peer-review process; in essence, the paper bypassed a level of scrutiny that's specifically tasked with uprooting poor-quality research. As discussed in *Chapter 4: Show Me the Research*, not all journals are created equal, and there's a broad range of outlets publishing everything from cream-of-the-crop research to science that's fundamentally flawed. A utopian world would see the media embody a sort of *last line of defense* to shield the public from low-quality research, but this is manifestly not the case. The public, therefore, have the responsibility for their own understanding of science.

Bad science does inevitably slip through the net, and it's seized upon by some individuals who'd welcome a complete dismemberment of the peer-review process. Such an action would favor poorly controlled, low-quality studies that serve to dilute the knowledge-base. Certainly, studies like Bohannon's highlight ways in which the process must evolve to force a more robust adherence to good standards of practice, and how journals must work to implement more widespread rigorous reviews and eradicate predatory pay-to-publish journals.

6.4.2. A Little Alcohol is Good for Your Health?

The second story concerns the notion that a glass of wine a day is beneficial to health, which has been propagated by hearsay and reinforced by the media for many years. Modest alcohol consumption was said to provide immunoprotective effects and reduce the risk of certain cardiovascular events. Wine was thought especially healthful due to the resveratrol and antioxidant content of the grapes. The claim results from a number of observational studies showing that light-to-moderate alcohol consumption evoked better cardiovascular outcomes and lower risk of mortality when compared to complete abstinence. There was an exuberant response to the advice because it condoned pre-existing behaviors that were once considered unhealthy. But it was a mistake to brashly accept the data and blindly comply with the resulting guidelines.

A large multinational report published in the *Journal of Studies on Alcohol and Drugs*[39] shed new light on the purported benefits of alcohol. The paper was a meta-analysis which utilized statistical techniques to combine findings from all of the suitable independent studies; it thereby increased statistical power and resolved ambiguous findings which would not be possible from an independent study. Meta-analyses are the most appropriate way of assessing the clinical effectiveness of a given intervention. The question posed by the authors was: *do moderate drinkers have a reduced all-cause risk of mortality [death] when compared to non-drinkers or heavy drinkers*? In other words, is there a *real* protective mechanism to the daily consumption of alcohol? Using data from 87 studies with a combined ~4 million participants, the researchers first plotted a graph of daily alcohol consumption versus relative risk of mortality, finding a J-shaped curve. This suggests that there is a slight risk of mortality associated with complete abstinence from alcohol, a reduced risk with low-volume drinking (1.3–24.9 g/day), and an exponential increase in risk with medium-, high-, and very-high-volume drinking.

According to the original model, those who completely abstained from alcohol exhibited a greater relative risk of mortality when compared to low-volume alcohol drinkers, thereby confirming the hypothesis that a glass of wine a day is beneficial to health. But the analysis was flawed. These are *observational studies* in which an assessment is made of the participant's behaviors, which are then discussed in the context of the outcomes. It's essential, therefore, that the studies have employed common methods, and have common characteristics in the way in which their participant groups are defined. To elucidate a clear understanding, therefore, the pertinent question is: *what criteria are used in assessing alcohol abstainers versus casual drinkers versus heavy drinkers, and what reasons might someone have for altogether abstaining from alcohol?*

The answer reveals insufficiencies in the interpretation. Many of the studies used in constructing the initial model included subjects with a history of alcohol abuse who had, more recently, chosen to abstain. In many cases, subjects abstained from alcohol either because they'd entered into an abstinence program, or because they were suffering the ill-effects of prior abuse. Some may also have abstained owing to a pre-existing medical condition which alcohol would have exacerbated. These confounding factors weren't controlled for in many of the initial studies. The inclusion of former drinkers in the abstainer reference group, therefore, artificially increased the relative risk, and falsely magnified the health benefits associated with low-volume drinking. Moreover, such definitional inconsistencies need only exist in a few studies to alter the group means, skew the data, and lead to erroneous results. When the data were adjusted to accommodate these covariates, the J-shaped curve diminished, evolving a chart showing little-to-no difference in relative risk between abstainers, occasional, and low-volume drinkers. So, low-volume drinking doesn't appear to be chronically harmful, but there's no protective effect as was first reported. Professor Tim Niami of Boston University Clinical Addiction Research & Education Unit, and one of the authors of the meta-analysis: *For the past 20 years many people have believed that 'moderate drinking' may be good for your heart, but research doesn't actually back this up.*

The take-home message is that accepting data at face-value, without further enquiry or independent analysis, can lead to false conclusions and the propagation of incorrect advice. Moreover, before advising others on causation between two variables (e.g., alcohol consumption and health outcomes), a clear mechanism should be proposed to explain the connection and minimize the likelihood of the *post hoc* fallacy, and this was conspicuously absent in the first model. Finally, consider that the initial advice remained unquestioned for decades, in part because it was a popular conclusion that was championed by the media. It conformed with a fallible human heuristic that there exists a single magical elixir that'll cure our ills and improve our health, even better if that elixir is something we enjoy. If one chooses to drink wine every day, then it should be for pleasure and not for some misplaced notion that it bestows health or is approved by medical science.

6.5. Chasing the Headline; the Time I was Offered Money to Bias Data

The topic of my master's research was carbohydrate use during endurance cycling; the pertinent finding was that sports drinks containing both glucose and fructose (as opposed to glucose alone) enhanced external carbohydrate use during exercise and improved performance in a cycling time-trial. This was novel research of the time and has since been established firmly in the literature. I'd scarcely submitted my thesis, which was published only in abstract form, when a colleague directed me to an article featured in the Daily Mirror newspaper (UK) which had cited my findings. The well-established performance physiologist Dr Miriam Stoppard OBE (not a performance physiologist) was responding to a letter written to her by a man wanting advice in preparing for his first marathon. In citing my research, Stoppard wrote that *researchers at the University of Hertfordshire suggest mixing traditional glucose drinks with fruit juices in order to maximise energy and run a faster marathon*. Stoppard studied medicine at Kings College, Durham, was research director, and won the title of *Journalist of The Year* in November 2008. Yet, to this day, it remains unfathomable to me how such an inaccurate interpretation could have stemmed from such a brief scientific abstract and adjoining press-release, neither of which made any mention of fruit juice or marathon running. One explanation is that Stoppard conflated *fructose* with fruit, the latter of which contains the former. But this tenuous link does the science of carbohydrate metabolism an injustice, not to mention the time and rigor channeled into the study. I contacted the paper to highlight the misappropriation, and also tried Stoppard herself on several occasions, but she was unavailable for comment. I still ruminate on how many runners may have suffered gastrointestinal distress in the marathon that year as a consequence of guzzling grotesque concoctions of fruit juice and glucose drinks on Stoppard's advice.

At a similar time, I was approached by a sports nutrition company who, having read the new and emerging studies in carbohydrate supplementation and glucose/fructose formulas, were keen to exploit the research and commercial appeal. As a prelude to an advert in a triathlon magazine, they asked me to conduct a small, independent study (in return for a fee) to test the ergogenic effects of their new supplement against that of their closest competitor. For the task, a professional cyclist had been recruited, and was to receive a fee by way of compensation for completing some time-trials in the lab. While quite meaningless in terms of making valid inferences, the sample size ($n = 1$) would be inconsequential amidst a backdrop of commercial science with a professional athlete. A young sports scientist, naïve and impetuous (and broke), I considered accepting the job. That was until they asked me to play a more considerable role than that of facilitator and assessor. During the second time-trial – that in which their product would be tested – they requested that I guide and motivate the subject, while ignoring him in the first trial during which he'd use the competitor's product. They even went so far as to ask that I verbally espouse the product's miraculous effects, thereby further reinforcing the subject's expectation bias; it was not a blinded trial by any means. In retrospect, I consider this a *fork in the road moment*; I'm pleased to say that, despite my inexperience, scientific integrity and honor prevailed and I declined to be involved. I believe the study progressed in my absence, but the ensuing ad-campaign received a vehement backlash.

Finally, despite the pessimistic picture I've been painting, it should be noted that not all media-reporting deserves to be reprimanded. There are some legitimate outlets which go to great lengths to probe beyond the attention-grabbing headlines to investigate the science at a deeper, more authentic level, and these outlets should be recognized. Indeed, there are a great many science journalists who are doing excellent work in accurately portraying new and emerging research. In general, these writers – reporting credibly on the science – exhibit two key traits that distinguish them from those who sensationalize the stories for superficial popularity: (1) intellectual integrity, and (2) critical-faculties coupled with scientific skepticism.

6.6. Organic Food

Organic food is that produced in farming systems that are free from man-made (synthetic) fertilizers, pesticides, growth regulators, or livestock feed additives. In recent years, there's been a dramatic surge in the sale of organic foods; in the US alone, revenue increased from $3.6 billion to $43.4 billion from 1997 to 2015[40]. In Chapter 3, I alluded to the notion that organic food is principally marketed on a logical fallacy called the *appeal to nature*, in which one asserts that a practice is necessarily *good* because it's *natural*. In this respect, the industry succeeds in exploiting our engrained bias for *natural* (non-synthetic) products. But if we're to integrate organic foods into a healthy-eating regimen, then we should do so on the basis of valid arguments, and not an informal fallacy of logic. So what reasons are there to eat organic?

First, the organic food industry purports to be more environmentally friendly and sustainable than conventional farming. While sustainable farming is a valid and worthwhile aspiration, the arguments are complex and controversial, and also outside the scope of this book. Lobbyists also claim, however, that organic produce is more *healthful* than food grown via conventional methods, and *that* claim most certainly *is* within the scope of this book. A survey presented at the annual meeting of the *American Agricultural Economics Association*[41] reported that 69% of consumers who regularly bought organic foods did so because they perceived it to be healthier. This is relevant because the perceived health-giving properties of organic foods unequivocally contribute to sales. Moreover, the claim appears to partly justify the higher costs, and is likely pertinent for exercisers interested in healthy-eating and athletes looking to augment their sporting performance. But the healthful nature of organic food is a fact-based claim that has been empirically tested.

A 2010 review of literature[42] synthesized 50 years of published studies (from 1958 to 2008), engaged in correspondence with subject experts, and hand-searched bibliographies. The authors concluded that *evidence is lacking for nutrition-related health effects that result from the consumption of organically produced foodstuffs*. The conclusion corroborates that of another systematic review published a year earlier by the same group[43], in which 55 studies deemed to be of appropriate quality were collated; this also concluded that *there is no evidence of a difference in nutrient quality between organically and conventionally produced foodstuffs*. The latter of these did suggest that organically grown produce may contain higher concentrations of certain nutrients, mainly phosphorus, but offered context by noting that phosphorus is widely available in the conventional diet, and that additional intake provides no benefit. Isolated studies show that certain organically grown fruits contain greater micronutrient concentrations than those grown non-organically, but such observations do not reflect scientific consensus. In other words, in order to harvest evidence (pun intended) that organic foods are more healthful, one has to cherry-pick the data (pun intended) to conform to a pre-existing agenda. By contrast, the bulk of the research, studied holistically, in its entirety, reveals no such pattern. Nutrient concentrations of fruit and vegetables also need to be corrected for size; i.e., organic produce is generally smaller than non-organic produce, and the *relative* nutrient concentration is likely to be artificially higher, but isn't congruent with a greater *absolute* nutrient quantity. A third systematic review from 2012[44] suggested that any differences in biomarker and nutrient concentrations between organic and non-organic foods were not significantly or clinically meaningful.

A common misconception surrounding organic food is that it's pesticide-free, but this isn't so. Pesticides are a necessary prerequisite to healthy crop growth. As the name would suggest, pesticides maximize crop productivity by controlling pests like weeds, insect infestations, and diseases. The label *pesticide-free* adorning organic food is a misnomer because organic crops are grown using *organic* pesticides (i.e., those derived from natural sources), as opposed to conventional farming which makes use of synthetic pesticides. Regulators only test for the latter, thereby permitting organic produce to circumvent the regulations. It's true that some data indicate a lower risk of contamination from detectable pesticides in organic foodstuffs; however, it's unlikely that there are differences in terms of the risk of exceeding the maximum allowed limits of exposure. Pesticide use is highly regulated by the government (for both organic and non-organic methods), and neither farming method yields produce that exceeds the safe upper-limit. And if you remain unconvinced, just washing your produce before use will help remove pesticide residue that remains post-harvest. But there remains an interesting twist to this story. An investigation in the 1990s[45] compared crop health between an organic pesticide (rotenone-pyrethrin) and a non-organic soft synthetic pesticide (Imidan). While two applications of Imidan produced the desired outcomes, an equivalent level of protection required seven applications of rotenone-pyrethrin. Environmental impact wasn't assessed, but it's plausible that the natural pesticide may have been more harmful given that it required more than thrice the number of applications. It's unfortunate that regulatory bodies don't screen for organic pesticides, as this may well impact on the sale of organic food.

Many eat organic because they perceive it to have a superior taste; while this is subjective, it may be the only valid reason. On two key aspects, the research is rather conclusive: (1) the poor evidence in favor of organic foods for being more healthful; and (2) the lack of any difference in health risk from organic versus non-organic pesticide exposure. There is, therefore, very little health-based justification for the greater expense of organic food. The fact that consumers still perceive organic foods as more healthful represents a victory for marketing over science, one that we must strive to overturn.

6.7. Fruit, Vegetables, and the Myths of Dietary Fructose

Fructose has been marginalized in recent years for its harmful effects, but this is mostly unjustified. That's not to say fructose is healthful; it's still a sugar, after all, and overconsumption can lead to weight-gain and insulin resistance. But the negative stigma surrounding fructose is also perhaps due to a common misconception in the public domain in which fructose is confused with high-fructose corn syrup (HFCS). The latter is a food additive, liquid at room temperature, comprising both fructose and glucose. These two single-molecule sugars aren't bound as in table sugar (sucrose), but exist in adjustable proportions depending on the desired sweetness of the food. Fructose sweetens the additive and is added to confectionary in high quantities. HFCS is used in a wide range of products including soft-drinks, cereals, biscuits, cakes, ice-cream, and table sauces like Ketchup (i.e., all the good stuff). Despite the controversy, however, there's nothing inherently *evil* about HFCS relative to, say, table sugar. Indeed, when the latter is consumed and meets saliva in the mouth, the chemical bond between glucose and fructose is broken by the amylase enzyme; thenceforth, it's chemically identical to HFCS. Both sucrose and HFCS are energy-dense and nutritionally meager, and one should abstain for the most part.

By contrast, fructose alone is a single-molecule, naturally occurring fruit sugar, contained in much of the fruit and vegetables we eat; the greatest amounts are found in apples, pears, grapes, watermelon, and various dried fruit (i.e., that which are most sweet). The fact that fructose is naturally occurring is inconsequential; arsenic is also naturally occurring, as are snakes. But the concerns around fructose – to the extent where some will advocate eliminating fruit from the diets of healthy people – are centered on the fact that fructose and glucose are both monosaccharides, simple sugars/single molecules. These differ from polysaccharides, like starch and fiber, which are many molecules long and, therefore, slower to digest in the gut. Our society is particularly wary of dietary sugar, and rightfully so. But the assumption that the digestive system treats both simple sugars in the same way is a false one.

So, how does your body differentially deal with glucose and fructose? Glucose is the *end-product* of carbohydrate breakdown. This means that, regardless of the carbohydrate ingested (be it fruit, pasta, or candy), the glucose is liberated to appear in the blood and get oxidized or stored. The various carbohydrate types differ in the *rate* at which glucose is released from food and enters the blood. Glucose is absorbed rapidly through the small intestine via a transporter, the name of which isn't important, but the transporter has a high-affinity for glucose and evokes a sharp rise in blood sugar and insulin concentrations. It's for this reason that glucose is designated the highest glycemic index rating of 100 and, indeed, all other foods are referenced against it. When you're contesting an endurance sport like marathon or triathlon, glucose is a valid means of augmenting performance owing to the fast rate of appearance in the blood, which affords the muscles a readily available source of external energy; this, in turn, spares internal carbohydrate stores for the latter stages of performance when fatigue is most likely. An equivalent effect isn't observed with more complex carbohydrates because, aside from giving you a stomach upset during a race, the energy would simply take too long to be liberated from its starchy prison. Glucose is rapidly removed from the blood by the hormone insulin and, depending on the conditions, will either be oxidized for energy, stored as carbohydrate in the muscle and liver, stored as fat on the body, or be excreted from the body in urine. Habitually high intakes of glucose are blamed, at least in part, for insulin resistance and type II diabetes. We should abstain as much as possible from taking glucose, although three exceptions during which it may be efficacious are: (1) during a bout of low blood sugar; (2) during an endurance race; and (3) during post-race recovery when carbohydrates likely need to be replenished.

By contrast, when consumed as a simple sugar, fructose is absorbed through the intestines via an alternative transporter which has a much lower transport capacity; it, therefore, results in a more *drip-appearance* in the blood. Once fructose enters the circulation from the intestine, it's shuttled to the liver and converted into glucose before finally being in a fit state to participate in energy metabolism. This is a time-consuming process; consequently, fructose doesn't evoke the same sharp rise in blood sugar concentrations as glucose, and has a far lower glycemic index of around 20.

Data indicate that a modest amount of fructose, when consumed simultaneously with a large amount of glucose (75 g), can actually improve glucose tolerance in healthy adults[46], which may have implications for those with poor glucose control. Indeed, the available evidence suggests that fructose may induce lower glucose and insulin responses after meals in patients with mild type II diabetes, when compared to most other carbohydrates[47]. Moreover, a systematic review and meta-analysis of 18 trials found that replacing dietary carbohydrate with fructose of a similar caloric content improved long-term glycemic control without affecting insulin in patients with diabetes[48]. If consumed in sensible (normal) amounts, there's little reason, therefore, to consider fructose harmful to the body. Unequivocally, one shouldn't conflate the physiological consequences of glucose ingestion with that of fructose, and I can assure you that only in the rarest of instances has obesity been caused by the excess consumption of fruit!

6.8. The 5-a-day Initiative

The World Health Organization (WHO) advocate that we should strive to consume a minimum of five different pieces of fruit and vegetables per day, which is about 400 g. These guidelines have been endorsed by government agencies the world-over, including the United States and United Kingdom, the latter of which introduced the program to the public via the Department of Health in 2002. The 5-a-day initiative is a media *hook* designed to catch the attention of the public; the scheme has its own logo (five green squares stacked in a diagonal line), it's used in the marketing of produce, used as a lynchpin in discussions on dietary health, publicized on television, in books, magazines, and booklets distributed by the government. The UK National Health Service (NHS) advises that a single portion should be:

> ...two or more small-sized, one piece of medium-sized or half a piece of large fresh fruit; or two broccoli spears or four heaped tablespoons of cooked kale, spinach, spring greens or green beans; or three heaped tablespoons of cooked vegetables; or three sticks of celery, a 5 cm piece of cucumber, one medium tomato or seven cherry tomatoes; or three or more heaped tablespoons of beans or pulses.

Such is the media emphasis on the initiative that we now associate 5-a-day with the theoretical optimum intake for healthy function. Unfortunately, the number is an arbitrary target conceived only for its modesty. Thus, 5-a-day is a realistic and achievable proposition for most members of society, but had the suggestion been 10–15 different pieces of fruit and vegetables each day, most people wouldn't have even entertained the notion. Despite the modest target, statistics suggest that approximately two-thirds of British adults fail in their bid to meet it[49], while in the US, only one in ten eats the recommended amounts of fruit and vegetables[50].

To further sour these reports, research from University College London[51] found that seven or more portions per day were associated with a positive effect on health and a reduced risk of disease. Moreover, they observed a *robust inverse association between fruit and vegetable consumption and mortality*; i.e., as consumption increased, mortality decreased. There was a general dose-response of fruit and vegetables on health, meaning that more was better, and the authors recommended aiming for closer to ten different portions of fruit and vegetables per day. Perhaps it's time to shift our thresholds.

6.9. Fad Diets

Obesity is a condition of abnormal or excessive fat accumulation, to the extent that health may be impaired[52]. It's clinically defined as a body-mass index (BMI; mass relative to height) greater than 30.0, although because BMI isn't able to distinguish between lean and fat mass, there exist other diagnostic metrics like waist-to-hip ratio and body-fat percentage. The human body has an (almost) endless storage capacity for subcutaneous body fat, hence obesity's widespread prevalence. The most recent statistics from the WHO estimate that one in three Americans are clinically obese, one in four Britains (although the latter are gaining on the former), and it's projected that by 2030 it'll be one in two Americans. Obesity is a worldwide issue, with North America, South America, Canada, Britain, some parts of Europe, Saudi Arabia, India, China, and Australia the countries most afflicted. Obesity is associated with an increased risk of cardiovascular disease, type II diabetes, and certain cancers. In the UK alone, obesity and related complications cost the NHS £6.4 billion per year[53].

With an increasing emphasis on tackling the escalating numbers of overweight and obese children and adults, many are becoming more health-aware and striving to change their eating and exercise habits. This positive motion has consequences because some lack the knowledge and understanding to implement these changes in a meaningful and constructive fashion. Driven by our heuristically wired brains, many of us turn to unproven gimmicks and crash weight-loss programs in an effort to fast-track success. These so-called *fad diets* are a victory for modern consumerism and the capitalist market. In his book *Sapiens*, Yuval Harari suggests that a societal ethic of indulgence has resulted in the global obesity epidemic which, in turn, fuels the economy because of overconsumption. It is also evident to me that, to curb growing obesity rates, people panic themselves into buying expensive organic foods or fad diets which they're convinced are the latest panaceas. We, therefore, contribute to economic growth twice over; first by overconsuming, and second by attempting to atone with commercialized eating regimens. Together, obesity and the subsequent use of unproven fads are a consequence of indulgence and naivete, respectively, two states of mind that can fill your stomach and empty your wallet.

In those unable to manage their weight in a more constructive manner, extreme dietary behaviors (fads) manifest as a vicious circle; the lack of a long-term benefit leads to fluctuating weight-loss and gain, with fad dieting repeatedly invoked to remedy the problem. The tenets of supply and demand are pertinent here; out of the fires of this public health epidemic, an entire industry has been forged to create and market fad diets. There are literally hundreds of commercial diets available, all offering rapid weight-loss and health, some with more legitimacy than others, and some that can be implemented safely as part of a healthy lifestyle. Here are a few examples of commercial diets with which you might be familiar, or have even tried: Fruitarian, Pescatarian, Intermittent-Fasting, Body For Life, Cookie, Hacker's, Weight Watchers, Atkins, Dukan, South Beach, Beverly Hills, Cabbage Soup, Grapefruit, Superfood, Israeli Army, Subway, Juice-Fast, Master Cleanse, Gluten-Free, Ketogenic, Liquid, Low-GI, High Protein, Low Protein, Mediterranean, Okinawan, Raw Food, Paleo, Slimming World, Smart For Life, and my personal favorite, the Inedia Diet (aka breatharianism) in which proponents claim the ability to live entirely without food and that *one can live on the Holy Breath alone*. Few really take breatharianism seriously, but other examples are less transparently absurd and manage to fake legitimacy with coherent logic and science-sounding claims. Even the ideologies underpinning mainstream practices like vegetarianism and veganism have been misappropriated and exploited to sell organized eating regimens, all of which make a claim to better health and weight-management, and packaged and sold to an eager market. There are too many fad diets to discuss independently, so I'll collate the bulk of the research generically. Before doing so, however, a few particularly contentious practices and ideas warrant closer scrutiny.

Superfoods are particularly pertinent to our discussion because the term has been popularized and propagated exclusively as a marketing tool for one reason: to sell product. Proponents suggest that superfoods have numerous health benefits, including protecting against cancers and heart disease, improving brain health (an ambiguous claim), improving the condition of the skin, and aiding weight-management. Around 100 foods have been labeled *super*, including blueberries, salmon, spinach, seaweed, broccoli, kale, Brazil nuts, walnuts, tomatoes, black raspberries, and soy. The term is repeatedly invoked by food gurus on television and social media to sell everything from smoothies and supplements, to recipe books. The people you don't find using the term are the professionals; doctors, dieticians, nutritionists, exercise scientists, and researchers, because the evidence doesn't justify the grand claims. Moreover, health professionals are concerned that the superfood marketing ploy might evoke negative health outcomes characterized by unbalanced diets that disproportionately favor certain foods, as well as extreme dietary behaviors.

There are two premises at work here on which the popularity of superfoods depends: (1) there exist certain foods with greater concentrations of micronutrients/antioxidants than others; (2) there are healthful, almost magical, properties of these foods that justify the greater intake. The first premise is fair, and the nutrient-contents of most foods do indeed vary, but there's no evidence of any special properties or healing powers to be acquired. Blueberries, for example, contain generous concentrations of the antioxidants vitamin C and vitamin K, but meager amounts of vitamin A. Kale has vitamin A and vitamin C aplenty, but paltry concentrations of vitamin D. Ultra-violet (B) rays from the sun, when striking the skin, interact with 7-dehydrocholesterol (7-DHC) and are converted to vitamin D, but sunshine provides no vitamins A, E, K, riboflavin, or pantothenic acid. Nutrient needs are, thus, fulfilled through a balanced diet comprising a range of fruits and vegetables (and a little sunlight). As is usual, the truth is comparatively mundane against the overstated and sensationalist marketing. Legislation came into effect in 2007 to ban use of the term *superfood* in the European Union, unless sufficient evidence could legitimize the claim. On this, we're still waiting.

The practice of juicing has also found popularity in recent years due, in part, to celebrity endorsements, television, and media coverage. Proponents claim that a juice-fast will improve energy and vitality, assist with weight-management, and facilitate body cleansing, but in reality, it's another poor substitute for a balanced diet. A typical juice-fast comprises between 800 and 1,200 kcal per day, from four to eight glasses of blended and strained fruit and vegetables. It's a tautology in sports nutrition that dietary protein is crucial for the growth and maintenance of muscle tissue, and that essential fats are needed for nutrient transport and cell repair. But a diet solely composed of fruit and vegetables omits these two important macronutrients and, even acutely, may lead to relative malnutrition. Given that juicing removes the fibrous pulp which slows intestinal absorption, blood glucose concentrations rise to a greater extent with juicing when compared to blending. For those individuals who cannot tolerate the inherent non-soluble fibers of fruits and vegetables, juicing may be a viable option when integrated into a balanced diet.

Smoothies, too, are conjoined with all manner of claims with no evidence to support the magical properties purported; there's no validity to a *brain booster smoothie*, or a *slimming smoothie*, or a *superfood smoothie*. While the benefits have been grossly exaggerated and extorted for their consumerism, smoothies (when taken in moderation) can be incorporated into a healthy diet and might prove beneficial for those with low micronutrient intakes. A recent study published in *Health and Education Behaviour*[54] assessed the impact of incorporating fruit smoothies into the breakfast regimens of two high-schools in Utah, Arizona. Following the ten-week intervention, the fraction of students consuming a full serving of whole-fruit had increased from 4.3% to 45.1%. Any intervention that fortifies micronutrient intake among school children is likely a positive one.

From the perspective of weight-loss alone, fad diets result in positive short-term effects. Dieters tend to lose weight during the early phase owing to the severe calorie restriction, and the necessity for the body to catabolize adipose tissue (fat) to maintain energy status. Even acutely, this may compromise both health and mood (lethargy and irritability are often reported), and diets are constrained by logistical practicalities and time-commitments. Nevertheless, be it juicing, superfoods, high-protein, low-protein, or intermittent-fasting, the *long-term* effects of fad diets on weight-loss are overwhelmingly negative. Most of the studies with reasonable follow-up assessments indicate that extreme calorie restriction is not long-term sustainable, and people tend to revert to old habits and reacquire all or most of weight lost during the initial purge. Certainly, there exist periodic success stories but, in general, the results aren't impressive, particularly when tested in the overweight or obese. Fad diets have poor long-term efficacy because they address the characteristics or symptoms of poor diet and weight-gain (i.e., excess calorie intake) but not the complex and multiple factors underpinning the condition, which might include: (1) the client's relationship with food; (2) the client's understanding of what constitutes healthy-eating; (3) the client's understanding of how to implement a long-term healthy-eating plan; and (4) the client's access to healthy-eating and physical activity. Rarely are these behaviors adequately addressed by such interventions in order to evoke long-term change. For chronic weight-management, sheer will-power and extreme dietary behaviors are insufficient. So, what approach *will* maximize the likelihood of long-term success?

As one would expect, there's no single solution but a complex series of interrelated mechanisms that'll have a different manifestation in each individual. What follows, however, are a few speculations. As always, education is important for some, but it's unlikely the critical factor. Most individuals have a reasonable appreciation of the daily foods they should and should not be eating, but if this isn't the case, one can learn the superficial basics of healthy-eating in a few weeks. The same principle applies to exercise. Beyond following a structured exercise training program, the emphasis must be on increasing physical-activity levels by *moving more* and increasing daily calorie expenditures. But this, too, isn't a novel conception. Certainly, accessibility is an issue for some and is a difficult problem to address, but interventions at a local (community) level have proven to be beneficial for many. Much time is squandered, however, by exercise practitioners arguing over which exercise modalities are the most effective at facilitating weight-loss. And while some modes and strategies may be more efficacious in this regard, there's mounting evidence that, for someone overweight, reductions in body fat will follow almost any intervention that increases overall activity-levels, including: daily walking, running, cycling, swimming, High-Intensity Interval Training (HIIT), resistance exercise, pilates, yoga, tennis, hockey, aqua-aerobics, etc. The key is to identify that which you enjoy and are likely to sustain in the long-term and, above all, is safe and appropriate for your individual case and is supported by your physician. But the foremost challenge we face in the long-term battle for healthy weight-management is convincing people to care in the first place; only then might they consider making the necessary changes.

Evidently, this is no minor obstacle, but it's unlikely to be overcome in its totality via the usual resources accessed by consumers, i.e., personal trainers, doctors, nutritionists, or dieticians. This is a problem that must be addressed through behavioral psychology interventions that help overweight and obese clients understand and change their relationships with food, as well as confront their negative behaviors. Only a finite group of practitioners are appropriately credentialed/qualified to implement such therapies. Until tackled at the root cause, fad diets and other symptom-treaters will likely remain a popular choice, despite their poor evidence-for-efficacy.

6.10. Detoxing

The term is ubiquitous with the health and fitness industry, and is so frequently employed in the media that it's become embedded in our everyday language. Just some of the detox protocols that claim to absolve you of a poor diet or an indulgent lifestyle include: juice-fasts, herbal supplements, herbal tea, homeopathy, coffee enemas (yes, really), ear candles, and footbaths. Detoxing – also known as body cleansing – is to engage in some form of extreme diet or ritual to purge the body of *toxins*.

Detoxes are attractive because they, too, conform to our desire for an easy, quick fix solution to a complex problem. But the notion of periodically abstaining from, or ridding the body of, *toxins* is characterized by a type of ritualistic purification, and so exploits inherent superstitions that have developed over many millennia. It draws inspiration from antiquated medical ideas that blamed evil humors – tiny unmeasurable demons in the body – for the cause of disease. It was an idea adopted by the ancient Greeks in the guise of the Four Humors, and it's a theme still favored by many alternative therapists. Still pervasive in many institutions to this day, subsects of the Catholic Church continue to perform exorcisms to rid the ill and infirm of demonic possession.

But such superstitions were conceived during a period of medical and scientific naiveté when there were scarcely any better explanations for ill-health. Advances in medical science, followed by the germ-theory of disease, have rendered the detoxing hypothesis obsolete. The word *toxin* assumes the same invisible, evil, undefined spiritual causes of illness that modern science has relegated to the middle-ages. The biological machinery responsible for removing toxic bodily compounds is well described. Scott Gavura at Science-Based Medicine comments:

> The colon remains ground zero for detox advocates. They argue that some sort of toxic sludge (sometimes called a mucoid plaque) is accumulating in the colon, making it a breeding ground for parasites, Candida (yeast) and other nastiness. Fortunately, science tells us otherwise: mucoid plaques and toxic sludge simply do not exist. It's a made-up idea

to sell detoxification treatments. Ask any gastroenterologist (who look inside colons for a living) if they've ever seen one. There isn't a single case that's been documented in the medical literature.

Before commenting on the research pertaining to detox diets, our first obligation is to consider its prior plausibility; *given our current understanding of nature and how things work, how likely is it that the claim(s) being made about this product are actually true?* Millions of years of positive evolutionary pressure have bestowed upon us extraordinary organic machinery that functions to detoxify the blood and eliminate waste; the liver and kidneys, and they do an outstanding job. The liver produces enzymes which process harmful substances in the blood to less harmful ones which are then dissolved in water and excreted via the renal system as urine. A healthy adult has everything he/she needs to neutralize most common toxic substances that accumulate daily. In just 36 hours, for example, the liver can render harmless a potentially fatal concentration of alcohol (although readers are discouraged from testing this hypothesis), although an obligatory knock-on effect on the immune system and recovery are to be expected. Detox practices offer nothing additional. Yet, despite the lack of prior plausibility, the hypothesis of detox has still been empirically tested.

In 2009, Voice of Young Science – a program that encourages early career researchers to play an active role in public science – released the *Detox Dossier*[55] in which they investigated the claims of 15 detox protocols. Products under scrutiny included footpads, detox water, dietary supplements, and hair straighteners (again; yes, really). The report concluded:

> No one we contacted was able to provide any evidence for their claims, or give a comprehensive definition of what they meant by 'detox'. We concluded that 'detox' as used in product marketing is a myth. Many of the claims about how the body works were wrong and some were even dangerous.

There's no credible evidence of positive outcomes associated with any detox protocol; moreover, none have been shown to improve performance, or even remove toxins. Another review[56] on the efficacy of detox diets concluded that the only studies observing positive results on toxin elimination or weight-management were *hampered by flawed methodologies and small sample sizes.* The popularized practice of coffee enemas has not been shown to detoxify compounds, or improve liver function, and there's no difference in antioxidant status in people treated with coffee enemas compared to those who consumed coffee in the traditional manner. More worryingly, *hot* coffee enemas that are administered rectally have resulted in predictable injuries[57]. High-dose vitamin injections have been assessed in a systematic review of studies[58]. In vitamin B_{12}-deficient adults, orally administered tablets improved hematological and neurological functions just as effectively as an intramuscular injection. Unsurprisingly, our digestive machinery appears perfectly well-suited to the task of extracting micronutrients from the food consumed. There are widespread reports of ill-health and side effects following detox diets, justified by the diet's proponents using an informal fallacy known as *special pleading*. Special pleading occurs when there's an attempt to cite something as an exception to a generally accepted rule or principle, without justifying the exception. For example, feeling energetic and revitalized during detox is attributed to toxins leaving the body. But feeling lethargic and irritable is explained via the same mechanism. Reasoning like this is unfalsifiable.

Given the lack of plausibility, the meager supporting evidence, the expense, the unsubstantiated claims, potential health risks, and detrimental effects on athletic performance owing to lethargy and potential malnutrition, the use of any commercial detox is actively discouraged. The premise of detox is rooted in several fallacious arguments, not to mention a misunderstanding of physiological processes. The displeasing nature of detox is that it promotes the notion that lifestyle indulgences should be punished, and that such indulgences can be compensated for with yet more extreme behavior. There are no quick fixes; if you're unkind to your body, the most effective means of recompense is to implement sensible but chronic lifestyle changes. Despite our foremost desires, a two-week juice-fast, liquid diet, or coffee enema will not immediately absolve you of dietary excess. The idea that it can, and hence the popularity of the detox, is rooted in an obsession with shortcuts. Quick fix detoxes are in direct opposition to efficacious practices like basic healthy-eating, increased physical activity, and regular exercise, i.e., practices requiring chronic changes and a degree of perseverance. But such practices are also free, and not packaged and commercialized. I find it strange that detoxes still win out, on many occasions.

6.11. The Irony of Ignorance

I have a love/hate relationship with health food stores. On the one hand, they're the arbiters and proponents of healthy-eating, encouraging greater consideration for food and lifestyle. On the other, it's disconcerting how these stores attempt to monopolize public health with the implication that we cannot be fit and healthy without pills, supplements, branded dehydrated kale, or vitamin-enriched sour-dough bread. It's certainly a strong business model, but one that depends on both consumer naiveté with respect to their nutritional needs, and the famous and fallacious *appeal to nature*. Consumers may be exploited more prominently in health and fitness because the very emotive subjects of aesthetics and longevity are at stake. And this is compounded by poor regulation that underpins the sale of product.

More recently, some of the larger health food stores have appointed in-house *experts* to advise customers on health foods and supplements. The uniforms they don are crisp and white, resembling nursing attire, thereby affording them a clinical and authoritative appearance. They bestow advice on all manner of topics, including: energy levels, immune function, sleep, recovery, stress, and depression. There's then a recommendation of supplements or herbal remedies to relieve a given set of symptoms.

It's a microenvironment in which *pseudo*science has the potential to thrive. Not to project malice onto the assistants, because there's probably little comprehension that they're selling ancient Chinese medicine, homeopathy, that an inscrutable minority of the supplements they sell have good efficacy, or that their attire is designed to exploit our reverence and trust for clinical practitioners. They too are pawns in a grand chess-game orchestrated by marketing teams the world-over.

It's axiomatic that businesses function to make money and, given that they have little obligation to actual science, they cannot be blamed for the manner in which they operate. Nor can the employees be held accountable because wholefood companies don't sell medicines, and they're not bound by the same rules and regulations as their pharmaceutical counterparts. They can, and do, sell whatever they like. It's an industry worth billions, but those being duped number far greater than those buying the product.

That's why if we're ever to transcend the world of *pseudo*-nutrition, a two-pronged approach is needed: (1) we must target the broad education of those buying the flawed products, and upskill them to judge, for themselves, if a product is worth the investment; and (2) we must target those who are being exploited into *selling* the product, because they are the ones in need of the greatest education; that is the irony of ignorance.

We've all seen the glossy magazine advert claiming six-pack abs in five-minutes a day, and the fad diets claiming rapid weight-loss, health, and vitality. Many have trodden those paths and stumbled; but where they have failed, we will succeed. We must implement critical-thinking to streamline our approach to healthy-eating, so that we utilize methods that work and discard those that only claim to. Skeptical skills are crucial in this regard, but the media must also be held accountable for perpetuating only the most sensational stories while simultaneously relegating real science to the cutting-room floor; the latter simply doesn't trend as effectively. In the next chapter, we progress organically to the multi-billion-dollar industry of dietary supplements and drugs. We discuss those that might work, those that won't, if and when to use them, and the legitimate risk of supplement contamination which has apparently taken by surprise the world of elite sport.

7

SUPPLEMENTS AND DRUGS

de·cep·tion.

The act of deceiving; deceive; fraud; betray; cheat.

7.1. Regulations? What Regulations?

It's believed that the ancient Romans drank the blood of fallen gladiators (as well as their urine) for its purported health benefits. This practice from an earlier civilization is reminiscent of even more antiquated traditions of European Stone Age tribes who thought that devouring a fallen enemy would somehow endow them with his power. While these practices appear to us as archaic, they're not exclusively archived in the histories of long-extinct civilizations; there are small factions which still believe that drinking urine will somehow bestow miraculous health benefits. But when it comes to the modern supplement industry, the mysticism and exaggerated claims borne of superstition and *pseudo*science are almost as prevalent today as they were in ancient Rome.

Despite recent changes in the way supplements are advertised, and the important work of the modern critical-thinking movement, the poor regulation was due in part to the Dietary Supplement Health and Education Act (DSHEA) of 1994, which dealt much damage to the stringent regulations of dietary and performance-enhancing supplements. Under the act, dietary supplements were listed on a special category of *foods*, and supplement manufacturers weren't required to seek approval from the FDA before marketing dietary supplements that were first produced in the US pre-1994 (as it's considered that such ingredients exhibit a history of safe use). Products developed since this time (termed *new dietary ingredients*, NDIs) must be reviewed (but not approved) by the

FDA. There are two caveats, however, that might eliminate the need for this secondary process: first, if the supplement contains *only dietary ingredients which have been present in the food supply as an article used for food in a form in which the food has not been chemically altered*; and second, if the company can demonstrate a history of use or other evidence of safety. The act has been criticized by, among others, Dr Steven Novella of Yale University:

> The deal that DSHEA and NCCAM made with the public was this: Let the supplement industry have free reign to market untested products with unsupported claims, and then we'll fund reliable studies to arm the public with scientific information so they can make good decisions for themselves. This "experiment" (really just a gift to the supplement industry) has been a dismal failure. The result has been an explosion of the supplement industry flooding the marketplace with useless products and false claims.

As discussed in Chapter 1, the FDA is laden with the task of proving empirically that the products are unsafe, rather than on the manufacturer to demonstrate safety. The bill, therefore, serves a multi-billion-dollar supplement industry rather than consumer interest. A commentary in the *New England Journal of Medicine* suggested:

> Since DSHEA became law, the number of available dietary supplements has skyrocketed from an estimated 4000 to more than 55,000. It is not known how many of the estimated 51,000 new supplements now on the market include novel (post-1994) ingredients, but the FDA has received adequate notification for only 170 new supplement ingredients since 1994 — undoubtedly a small fraction of the ingredients for which safety data should have been submitted. Indeed, both the industry and the FDA acknowledge that many new products have been introduced without any assessment of safety.

The aforementioned is a shocking indictment of a flawed and broken system. More recent guidelines implemented by the FDA for the testing of supplement safety include detail on the documented history of use, formulation and proposed daily dosage, and the recommended duration of use (e.g., intermittent or long-term), but many believe that this remains insufficient, and loopholes in the guidelines leave manufacturers with extensive freedoms. Indeed, while there appear to be modest sanctions in place to increase the likelihood of safety with respect to dietary supplement ingredients (pre-1994 history of safe use, and a post-1994 FDA review), companies can pronounce ingredients as *self-affirmed generally recognized as safe* (GRAS), whereby an in-house team of scientists evaluate the safety of their ingredients, without any requirement to submit documentation to the FDA, although the evidence of safety must be held on record in the instance the product is retroactively challenged. This process allows companies to avoid FDA applications for NDIs.

All these policies and regulations prioritize supplement safety, but speak nothing to product effectiveness. In 1999, the U.S. Court of Appeals for the District of Columbia Circuit ruled that qualified health claims may be made and approved, so long as the statements are truthful and based on adequate science. Research relating to claims can, therefore, be submitted to the FDA/FTC for approval. The scientific evidence is required to have been collected in an *objective manner* by persons *qualified to do so, using procedures generally accepted in the profession to yield accurate and reliable results*. This involves at least two clinical trials that support the given structure/function claim. Nevertheless, the quality and robustness required of these reports are ambiguous and still evolving, and what would satisfy one scientist as *scientifically robust* may not satisfy another. Instead, the data in such a report should have to satisfy an *absolute* standard of rigor. There are further insufficiencies to the policy. It makes no account for instances in which companies strongly imply claims without overtly stating them, thus allowing them to effectively circumvent the regulations. Moreover, to address sanctions, many established supplement companies now employ research and development directors and task them with conducting the studies and producing the scientific reports necessary to pacify the regulating bodies. Regardless of one's moral and ethical standards, conducting research to retroactively address a pre-determined claim is bad science.

With around 50,000 dietary supplements registered with the Dietary Supplement Label Database[4], such convoluted and ambiguous regulations are ill-befitting of the reach and impact of the market. It's become a proverbial minefield just discerning those that are effective and safe and those that aren't. It falls to the consumer, therefore, to challenge those facets of the industry that consider evidence-for-efficacy a burden rather than a responsibility. Accordingly, we'll now scrutinize more closely the aspects of the supplement industry that might specifically influence our buying decisions, by first answering the foremost question.

7.2. What are Supplements?

I labored over the chapter title because, when you consider the aforementioned question, it's apparent that there isn't a lucid answer. Some journals and governing bodies have attempted a definition, which I'll address, and I'll define the term as I intend on using it. Moreover, if we're to think clearly about our use of such products, it's necessary to first challenge our commonly held notions pertaining to the supplement industry. The European Food Safety Authority (EFSA) define supplements as *concentrated sources of nutrients (i.e., minerals and vitamins) or other substances with a nutritional or physiological effect that are marketed in 'dose' form (e.g., pills, tablets, capsules, liquids in measured doses).* This definition alone doesn't distinguish between a *nutritional* and a *physiological* effect. Rationally speaking, all bodily effects are physiological, and supplements may just be a nutritional means of manipulating physiological function. Nevertheless, *supplements* could be labeled more broadly as anything additional to the diet that may serve to improve health and/or performance (or at least thought to do so). We generally take supplements to fortify our diet with a compound not habitually consumed. The term is somewhat generic and it blurs a number of boundaries. For example, from a dietary perspective, whey protein is considered a supplement because it offers a nutritional value. Caffeine, by contrast, doesn't offer a nutritional value but is regarded as an ergogenic supplement by performance nutritionists who might recommend it to stimulate alertness. In this context, all legal performance-enhancing drugs are supplements, but clearly not all supplements are drugs.

A recently update review published by the *International Society of Sports Nutrition (ISSN)*[4] classifies dietary supplements as those taken orally (see Chapter 1). I think theirs is a lucid definition, but not the one that necessarily covers all eventualities. In some rare instances, an individual may benefit from regular use of a multivitamin (i.e., if they're shown to be clinically deficient in one or more micronutrients). As alluded to above, a physician may choose to treat anemia (diagnosed following a blood test) with a course of vitamin D – taken orally or injected – because vitamin D might be considered necessary to *supplement* the diet. But are both oral and injectable forms considered a dietary supplement? It's surely addressing a dietary insufficiency, but according to the ISSN statement, a prerequisite is that the dietary supplement be intended for oral ingestion. Now further consider powdered whey protein; it's axiomatic that this be labeled a nutritional supplement because we're *supplementing* our diet with protein to better meet our nutrient demands (which tend to increase with hard training). What if you, instead, decided to take more protein simply by eating more high-protein food (e.g., chicken or eggs); is the additional chicken now considered a supplement? It could be considered supplementary, but most would reject such a notion. But removing the abstractions we've conjured surrounding powdered food, beyond our preconceived bias toward the fact that one is *natural* and one is not, is there really a quantitative difference between protein from chicken breast and that from protein powder? Both chicken breast and whey protein evoke similar outcomes when ingested. What if the chicken breast were dehydrated, rendered completely devoid of water, and ground into a powder to be taken *ad libitum*; would it then be considered a *supplement*? Put caffeine in the same light; if it's a performance-enhancing or ergogenic *supplement* when taken as a pill (which is a fair premise), what about caffeine taken as liquid espresso (for the same purpose, at the same timepoint, to induce an identical physiological response). Is one mode of ingestion (liquid, pill, or gum) considered more of a supplement than another, given that the outcome (with the exception of the absorption rate) will be similar? Am I ordering supplements every time I visit my local coffee shop? Is the form in which the substance is administered really sufficient to cement its categorization? This is surely just semantics.

Some illegal drugs like the male sex hormone testosterone – banned in athletic competition – can be taken orally, transdermally using a patch on the skin, or injected; so how is this to be classified? Consider further that, despite there being a gulf between the two in terms of the physiological outcome, both caffeine and testosterone are drugs taken to enhance performance. Other than caffeine's legal status, what makes the former a supplement and the latter a drug? There was a time when the World Anti-Doping Agency (WADA) listed caffeine as a banned substance in high concentrations due to rightful concerns about the risk of overdose. But we'd scarcely label something as a supplement on the basis of its legality.

While many of these hypotheticals aren't to be met with concise answers, for the sake of this chapter, I'll settle for the imperfect classification of *supplements* as those substances that can generally be obtained from food (e.g., protein, creatine, caffeine, etc.), and *drugs* as substances that influence physiological function and that aren't normally obtained through the diet (e.g., testosterone).

7.3. Are Supplements Safe?

Few would disagree that the degree of safety depends on the supplement. Any ingredient consumed with sufficient vehemence can have harmful effects on the body; nevertheless, the consequences of excess caffeine are likely substantial relative to excess multivitamins. While caffeine is now legal according to WADA, the National Collegiate Athletic Association (NCAA) will suspend any athlete with urinary caffeine concentrations above 15 mcg/mL, which is equivalent to around eight cups of brewed coffee. This would likely result in blood caffeine concentrations of around 500–800 mg. According to the MAYO clinic, this amount is likely to be safe for most healthy adults, despite such values approaching the safe upper-limits, and individual sensitivity dictates tolerability. There's little research on the negative effects of a long-term high-protein diet, and while rodent studies suggest that there may be chronic changes in kidney function and/or an increased risk of kidney stones, these findings haven't been replicated in humans, and there's no evidence denoting such changes as inherently harmful. Moreover, there's a quantitative difference between diets that are *high* in protein and those that are *excessively* high in protein; the former tends to reflect amounts suggested in the published guidelines. In any case, lax regulations mean that products are rarely withdrawn from the market until there are demonstrable negative effects, and in that sense we're all guinea-pigs for the supplement industry. Ephedra, for example, is a potent stimulant that was once sold freely, and legally. It was withdrawn from circulation (and regulated as a drug) following cases of ephedra poisoning which reached 10,326 in 2002, with over a hundred people needing critical-care hospitalization. Seven people died from using the drug in 2004. Removing ephedra from the market took eight years, and was initially overturned by the industry in 2005, although ultimately upheld by the U.S. Court of Appeals in 2006. There are many ergogenic aids that are banned in professional sport either because they provide the user with an unfair competitive advantage and violate the spirit of fair-play, or more often because the supplement isn't considered safe for human consumption; the latter condition usually refers to drugs like stimulants or steroids. But you'll surely avoid the dangers of banned substances, and the consequential positive drugs test, just so long as you abstain from illicit drugs and deal exclusively in legal supplements, right?

7.4. Drugs in Your Supplements

Wrong. There's now sufficient data to suggest that the contamination of legal supplements with illicit drugs isn't just possible, it's common. In a former appointment, I was tasked with delivering nutrition workshops to professional sports teams comprising athletes and coaches, in which the discussion of supplement use was a priority. In my preparation for an upcoming session with a rugby club, I'd unearthed an alarming but pertinent statistic; of the 44 athletes at the time serving competitive suspensions in the UK for positive drugs tests, 50% (22 of 44) were competing in Rugby Union. Moreover, most of those athletes were academy players, i.e., under 18 years of age. This is an ominous statistic, but one not altogether surprising. Players transition from the junior to senior squads at the age of 18 years. This is likely an unnerving prospect for young athletes, with enormous pressure to be bigger, stronger, and more robust such that they might stand a chance of competing alongside their more developed senior peers. This doesn't excuse the use of illicit performance-enhancing supplements or drugs, but it might explain it. A culture shift within the sport is needed, and should align with broader education on the dangers of steroid use.

But several of these doping instances, along with many more on the world stage, were attributed to legal supplements (e.g., mass gainers and whey protein) that were contaminated with illegal drugs, thereby incurring the wrath of anti-doping officers. The magnitude and prevalence of this problem were examined, in 2004, by a team of researchers from the Institute of Biochemistry, German Sport University, who conducted a large investigation on the international supplement market[59]. They purchased 634 non-hormone-containing supplements, from 215 different suppliers, across 13 countries. A majority were purchased over the counter in stores, but others were shipped via online retailers. After isolation from the supplement matrix, the researchers used a technique called gas-chromatography/mass spectrometry (which basically vaporizes and then separates the sample into its various constituents) to test for the presence of prohormones and anabolic steroids. The results were surprising; of the 634 samples analyzed by the group, 94 of them (15%) were found to contain anabolic-androgenic steroids that weren't declared on the product label. Most of these *positive supplements* came from companies based in only five countries: America, the Netherlands, Great Britain, Italy, and Germany. Finally, and perhaps most importantly, many of the contaminated supplements contained prohormones in concentrations sufficient to trigger a positive drugs test. In 2005, micronutrient supplements including vitamin C, multivitamins, and magnesium were confiscated owing to cross-contamination with stanozolol and metandienone (anabolic/androgenic steroids)[60].

Presently, there's a growing list of professional athletes serving lengthy suspensions for positive tests that they attribute to a contaminated supplement. But whether a positive test results from deliberate misconduct, or from inadvertently taking a contaminated supplement, there's an equivalent outcome: a competition ban and a tainted career. While it might seem improper that both circumstances be met with similar sanctions, it's necessarily so because the inadvertent nature of the offense cannot be proven. Therefore, athletes must take responsibility for the substances they use, and they should work only with trusted professionals. There exist comprehensive resources to aid both athlete and coach in smart decision-making regarding supplement use. For example, subsequent to the 2004 study, the Cologne List® was developed; an initiative compiling lists of nutritional supplements that present a minimized risk of contamination from banned substances. Moreover, from www.informed-sport.com:

> Informed-Sport is a global quality assurance program for sports nutrition products, suppliers to the sports nutrition industry, and supplement manufacturing facilities. The program certifies that every batch of a supplement product and/or raw material that bears the Informed-Sport logo has been tested for banned substances by LGC's world-class sports anti-doping laboratory.

The FDA recently published online a list of products marketed as *dietary supplements* that were tainted with hidden ingredients deemed *hazardous.* While an enormous 923 products were listed, the FDA cautioned that it *only includes a small fraction of the potentially hazardous products with hidden ingredients marketed to consumers on the internet and in retail establishments. FDA is unable to test and identify all products marketed as dietary supplements on the market that have potentially harmful hidden ingredients. Even if a product is not included in this list, consumers should exercise caution before using certain products.* The advice is sensible and measured. How disconcerting that this appears to be the tip of the iceberg with respect to the full range of potential hidden ingredients in over-the-counter dietary supplements. Of the list provided, 349 were advertised for weight loss, 90 for muscle building, and 462 for sexual enhancement, with the remainder for *other functions*.

What reasons are there for such frequent and widespread contamination? Many such supplements are manufactured by companies that also make prohormones, and likely in the same factories; accordingly, inadvertent contamination is possible. But the Cologne study[59] exhibited many contaminated supplements from companies that didn't independently manufacture prohormones. It's plausible, therefore, that some companies *deliberately* sneak illegal prohormones into their supplements. This would enhance the product's potency which would, in turn, increase sales. It's a heartless business, especially when considering the numerous amateur and professional athletes whose careers are put in jeopardy from use of such supplements. Be cognizant that the majority of supplement users are recreational exercisers – gym-goers, club runners, triathletes, bodybuilders – who remain unburdened by periodic visits from the anti-doping drugs testers. Would you begrudge a tainted supplement if it enhanced your outcomes, particularly if you could claim ignorance to its composition?

7.5. Should I Ever Use Supplements?

So, when might it be an appropriate time to integrate supplements with the diet, and how do we ensure that only those that are safe and that have sufficient efficacy are chosen? Most nutritionists and dietitians agree that a food-first approach is best, be it for an athlete or an exerciser. This means prioritizing your diet and contriving to optimize your nutritional intake before pushing the envelope. Too many people bypass this crucial step to exploit the benefits of supplements before they're in a position to maximize the effects. A food-first approach is preferred but not because it's *natural*; such ambiguous conceptualizations were challenged earlier. It's more related to the large gains one can make by optimizing diet, compared to the relatively small gains likely with supplement use. Even those products that work (with strong evidence-for-efficacy) will be squandered on someone whose diet is wholly and chronically insufficient for their needs.

The food pyramid was an initiative conceived several decades ago, featuring the major food groups (grains, fruit and vegetables, meat, and dairy) organized in a hierarchical pyramid in the amounts recommended for a *healthy diet*. The early incarnations were criticized for several reasons, including a general objection to a hierarchical structure, and the implication that grains should be the foremost component of a diet. The pyramid had several manifestations over the years, with the most recent from the US Department of Agriculture in which it takes the form of a plate divided into five sections for each of the major food groups (see Figure 7.1). While the new icon received a more positive reception, there are inherent problems with the plate, including the lack of individualization for a population with broad needs (consider a sedentary 60-year-old versus an athletic 20-year-old). There's also a lack of provision for vegetarians and vegans, the latter of whom also refrain from dairy and may need guidance with respect to valid sources of protein. Nevertheless, the food plate wasn't designed to serve as a comprehensive dietary strategy, but rather a simple and accessible *how to* for making sensible food choices. It can, therefore, be used to structure our discussion on the most impactful nutritional adjustments for health and performance, and whether sufficient space can be made for a side of supplements. Note that below isn't a definitive doctrine, but rather some suggestions on how to organize diet and supplements into something resembling a hierarchical structure; it's also a comment on how the importance of food likely supersedes that of supplements in the healthy diet.

Grains have replaced the generic *carbohydrate* group, but first consider the macronutrients (protein, carbohydrate, and fat) holistically. Learning how to optimize your macronutrient intake will arguably provide you with the greatest outcomes whatever your goals (weight loss, muscle mass, endurance performance, etc.). If exercising regularly, especially at high intensities, then carbohydrates will serve an increasingly valuable role in your diet. A chronic lack of carbohydrate increases circulating stress hormones during exercise, and can suppress immune function[61]. But knowing how to strategically manipulate carbohydrate, through combined knowledge and experience, results in cellular adaptations that augment fat-burning capacity and enhance endurance performance[62]. Protein's essential role in the growth and maintenance of muscle tissue is well-established, but it's notably more important for those engaged in strenuous training and for whom the recommended daily amounts increase. Protein is sourced from meat, fish, poultry, dairy, and non-animal plant sources. The decree is that macronutrients can influence health status and exercise performance profoundly, and one should first endeavor to optimize this component of the diet.

Next, there are fruits and vegetables; while both are a source of carbohydrate, the latter isn't a particularly good source. Both are rich in the vitamins and minerals essential for healthy function. These micronutrients are crucial for the repair of bone and skin, maintaining a healthy immune system (to stave off viruses, for example), repairing cellular damage, and helping unlock energy from the food consumed. Envision laboring after your fitness for many months, adhering to a carefully planned and periodized program, only to neglect your diet and suffer a micronutrient deficiency which afflicts you with the flu. Proceeding to spend the majority of the next two weeks prone on your sofa, dispelling the contents of your nose, will deal a severe blow to your hard-won fitness. As aforementioned, 5-a-day is generally insufficient for young, active individuals, and grossly so for athletes in whom at least 10-a-day should be a legitimate target. The integration of fruit and vegetables into the dietary regimen should be considered with equal enthusiasm as one considers macronutrients.

Finally, assess hydration status (although surely for many this will need to happen first). The blood is a fluid; in hot conditions and during exercise, one secretes sweat through pores in the skin. Sweat evaporates and liberates heat from the body, thereby helping to maintain core temperature within a satisfactory range. That sweat comes primarily from the blood (although not directly, it's first moved into the interstitial spaces); accordingly, sweating causes dehydration and a decrease in blood volume. With less blood circulating the body, the heart must pump with greater frequency to maintain the overall volume being delivered to the muscles and organs. Such a decrease in reserve is called *cardiovascular drift* and it makes exercise harder. The take-home message is that dehydration dramatically influences exercise capacity, as well as energy levels and cognitive function. Health and performance depend greatly on adequate hydration, and there exist easy and accessible methods of monitoring hydration like the urine color chart. For most people, the thirst response is sufficiently sensitive to remain hydrated throughout the day, although a particularly busy schedule may see you ignoring your subtle thirst response until you've been rendered more substantially dehydrated. Moreover, the thirst response might be a poor indication of hydration when exercising or when the ambient temperature increases.

When these interrelated components of the diet (intake of carbohydrate, fats, proteins, micronutrients, hydration) have been carefully considered and optimized, then it may be reasonable to invoke supplementation to address any outstanding needs. Even then, careful research should predicate your decision to ensure only those supplements with a sufficient empirical base are considered. Each should be trialed individually so that any differences (perceptual or

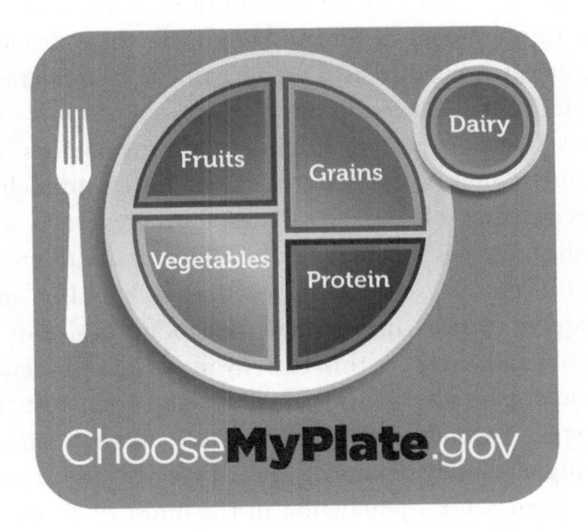

FIGURE 7.1 The US Department of Agriculture (USDA) food plate, divided into five sections for each of the major food groups.

physical) can be discerned. If you're a competitive athlete, then it's critical that this process is managed through your coach and an appropriately qualified nutrition expert, particularly given the high prevalence of supplement contamination. It's prudent, therefore, to use only those supplements that have been tested by third-party organizations. In the US, the United States Anti-Doping Agency (USADA) recognizes NSF Certified for Sport® as the program best-suited to reduce the risk of using a tainted supplement. Moreover, Informed-Sport is a global quality assurance program for sports nutrition products, suppliers to the sports nutrition industry, and supplement manufacturing facilities. Presently, there's little tolerance for athletes who test positive for either illicit drugs or contaminated supplements, given that they amount to the same thing. The responsibility of using safe and effective supplements falls firmly on the shoulders of the consumer.

7.6. Fat-Burning Supplements

Of all the reasons underpinning supplement use, increasing the body's capacity to metabolize fat is surely among the most cited. Such an outcome is sought for the purpose of enhancing endurance performance (improving fat oxidative capacity is a keen goal of the marathon and ultra-marathon runner), or perhaps instead to decrease body fat for health and/or aesthetic reasons. The claims of the health and fitness industry tend to focus on the latter objective for obvious reasons, and those supplements purporting to aid in the process deserve scrutiny. Of those aimed at enhancing fat metabolic pathways, L-carnitine is one of the most widely produced and distributed, and yet a majority of the associated products are based on traditional research which suggests that it's scarcely possible to enhance muscle carnitine content in humans, resulting in minimal (if any) change in physiological function. As such, these products are at best monetizing a concept, a hypothesis, a premise that until quite recently failed to be accepted under controlled conditions.

Carnitine is a transport protein in muscle cell membranes, and it's largely responsible for the transport of free fatty acids to muscle mitochondria[63]; i.e., carnitine has an essential role in fat oxidation during exercise. The availability of carnitine in the muscle is a major rate-limiting step in this process. The early premise was that if muscle carnitine concentrations could be enhanced, there may be a consequent increase in fatty acid oxidation. As already alluded, there would be two obvious benefits to this: first, by increasing fat metabolism during exercise, one could spare internal carbohydrate stores and delay fatigue during endurance exercise. Second, aside from aesthetic implications, a supplement capable of shifting metabolic flexibility toward greater fat-burning would have profound consequences for a population in the midst of an obesity epidemic. Unfortunately, as is often observed with an ambitious premise, the evidence is found wanting. A 2018 sports nutrition and supplement review[4] concluded that, to date, *the majority of the data continues to suggest that carnitine supplementation does not markedly affect muscle carnitine content, fat metabolism, exercise performance, or weight loss in overweight, obese, or trained subjects.* The comprehensive review cites individual studies too numerous to reference here, but further reading is encouraged. Yet, L-carnitine is still available in powders, pills, drinks, and energy bars around the world. Could it be that the premise alone was sufficient on which to found a multi-million-dollar dietary supplement?

Studies did eventually discover a means by which muscle carnitine content might be augmented, but with several costly stipulations. Researchers demonstrated in 2006 that in the presence of high insulin concentrations (hyperinsulinemia), total muscle carnitine content was bolstered by as much as 13% (from 22.0 to 24.7 mmol/kg/dm), which was deemed statistically significant[64]. A follow-up study from the same group supported the initial findings using carbohydrate ingestion to stimulate insulin secretion[65]. Although on first impression these findings appear to have clinical implications in the treatment of diabetes and obesity, and perhaps also for endurance performance, increasing muscle carnitine content by this method necessitates high insulin concentrations actioned either via direct infusion of insulin into the blood, or naturally via the concurrent ingestion of fairly high amounts of carbohydrate (which, in turn, stimulates the obligatory state of hyperinsulinemia). Supplementation with L-carnitine would appear counterproductive, therefore, for obese or overweight individuals in whom caloric intake would need to be restricted, although it may be less of a concern for competitive athletes with a high energy turnover. The research is compelling, but further investigations must predicate any practical recommendations.

As with all commercial products, particularly supplements, we must be wary of deceptive marketing strategies designed to mislead the consumer into inferring dramatic performance gains in light of only modest evidence. An established sports nutrition company, until recently, advertised an electrolyte supplement which evoked a 40% increase in fat oxidation during exercise. Ostensibly, this is a colossal advantage, and one which was demonstrated under controlled laboratory conditions. But the product – a calorie-free electrolyte formula – was tested against a standard 6% carbohydrate solution. It stands to reason that the non-calorie supplement would induce greater fat oxidation, since its comparator was designed specifically for the purpose of enhancing carbohydrate use. Indeed, in the absence of carbohydrate to burn, any individual with healthy digestive/metabolic function will oxidize more fat with a sugar-free solution. It's analogous to a new running shoe marketed proudly as being able to improve running performance by 30% relative to a competitor's brand, when the latter is a rugged hiking boot; the products are conceived for different purposes. Had the electrolyte product been tested against water, the rate of fat oxidation would likely have been comparable. The advert isn't false or fallacious in its claim, but nor is it appropriately candid and transparent regarding the conditions of the test.

In this domain of sports supplementation, marketing is king, and when research is uncharacteristically cited, it usually warrants closer scrutiny.

There are numerous other commercially available non-carnitine-containing products which also stake a claim at improving fatty acid transport and/or oxidation, a full discussion of which is outside the scope of this book. Briefly, conjugated linoleic acid (CLA), caffeine, and green tea have received attention in the literature, and while data on green tea are interesting, it's not sufficiently robust so that firm recommendations can be made. There are several drugs that may have a strong influence on metabolic function and substrate utilization. Sympathomimetic amines like ephedrine or clenbuterol have been used (illegally) by athletes in professional sport, the latter famously by Alberto Contador who was banned for two years and stripped of his Tour De France title in 2010 after testing positive for the drug. Classified under the umbrella term *stimulants* and *thermogens*, few of these illegal drugs are considered safe, and are judiciously confined to the WADA list of banned substances. Thus, the list of supplements claiming to positively affect fat metabolism is both substantial and commercially driven, and their popularity in no way reflects their scientific validity.

7.7. Protein, Protein, Everywhere

Another of the most keenly sought-after commercial supplements is protein, perhaps justifiably in this instance. The muscle's structural and functional proteins are in a constant state flux; that is, proteins are forever being broken down and reassembled. While dietary protein is important for all individuals, regardless of activity level, we see greater protein turnover following hard training and so protein requirements are higher in exercisers and athletes. Amino acids (the building-blocks of protein) also regulate immune function. Accordingly, meeting your individual protein demands is an important component of a healthy diet. While the evidence is in favor of dietary protein as a potent stimulator of muscle protein synthesis (to aid in recovery), there's much confusion and misinformation with respect to what protein is and isn't. Protein isn't a magical anabolic compound that'll bestow upon you muscles without the need of exercise. The supplement form isn't preferable to the dietary form (that derived from food), and supplements provide nothing that cannot be fashioned from an amino-acid-rich source of dietary protein, like chicken, eggs, or fish. There's likely no requirement to take protein during a training session so long as it's consumed regularly throughout the day. There's a small dose-response to protein up to a very specific threshold which we'll discuss below, but more isn't necessarily better. Protein doesn't burn fat and, despite the myths and hearsay, surplus protein doesn't simply escape the body in urine; there are very clear metabolic pathways that convert excess amino acids to fat. As an aside, detraining doesn't degrade muscle to fat, nor does fat transmogrify to muscle upon beginning a new weights program. You *can* meet your protein needs through a vegetarian or vegan diet, although it's generally more laborious and warrants careful planning.

Much of the confusion regarding dietary protein and supplements pertain to how much is needed per day and per dose. Fortunately, this is one topic on which there's a clear scientific consensus. What follows is an abridged overview, although it's suggested that you examine the data more closely for targeted guidelines. Daily protein targets for sedentary people (those not physically active) are 0.8 g of protein per kilogram of body mass (0.4 g per pound), although recent suggestions are that it may be slightly higher than this. By way of example, an individual weighing 80 kg (176 lbs) would need a daily intake of ~70 g. Those training but not seriously should aim for intakes between 1.0 and 1.5 g per kilogram of body mass (0.5–0.7 g per pound), and those with more ambitious sporting goals with an emphasis on strength, power, or ultra-endurance should target intakes between 1.5 and 2.0 g per kilogram (0.7–0.9 g per pound). Amounts exceeding 2.0 g per kilogram per day are unlikely to afford additive effects on muscle protein synthesis, although values up to 2.5 g per kilogram per day might be appropriate in people with a very high calorie turnover. With respect to optimal amounts per dose, a 20 g bolus appears sufficient to maximize muscle protein synthesis[66], with up to 30 g appropriate for larger individuals (>85 kg or 190 lbs). Protein needs may also be higher in older adults[67]. With respect to timing, a protein feeding strategy of around 20 g every 3 waking hours is more effective at stimulating muscle protein synthesis than feeding ~10 g every 1.5 hours, or bolus-feeding of ~40 g every 6 hours[68]. Finally, protein ingestion before sleep appears to be an effective strategy to increase muscle protein synthesis rates overnight[69]. And so goes our current state of knowledge regarding some of the key aspects of protein intake. The evidence is fairly refined and unambiguous, and while there are contentions over some of the finer details, these broad guidelines will suffice for most people. It's just unfortunate that commercial or anecdotal advice so frequently contravenes the science.

As suggested, obtaining protein with the full complement of amino acids may prove problematic for vegetarians and vegans, because non-animal sources of protein generally have an inferior range of amino acids. Nevertheless, with a little planning and ingenuity (e.g., by blending those plant foods with complementary amino acid profiles), one can usually fulfill their requirements through dietary means. Collectively, even if protein needs are considered higher than normal, expensive supplements are not a prerequisite.

Protein has emerged as a tremendous commercial powerhouse, and not just that which is sold in supplement form; protein is now synonymous with health. Nuts, seeds, tofu, and bulgur wheat (among others) are labeled distinctly as *a source of protein* even though the actual content is rather paltry. We can buy cereals, cereal bars, and oatmeal that are *packed with protein*, high-protein yogurts, high-protein bread, high-protein pasta, and high-protein chocolate milkshakes. While this serves us in one capacity (high-protein yogurts are sensational), the hype inclines many to a relative protein glut, overfilling their shopping baskets with high-protein produce they don't need. Protein isn't a magical macronutrient; it contains no special healing-powers, it won't make you slim, endow you with hulking muscles, or improve your fitness. The *ad nauseam* use of the word serves to diminish its potency, and consumers must pay closer scrutiny to the protein content of foods to ensure the nutritional information on the reverse of the packet doesn't contradict the slogan on the front. If protein demands increase through training, then protein intake should increase congruently, but excess won't serve you. Study the science, read the papers, follow the guidelines, and this modest macronutrient will play its partial role in your healthy-eating regimen and facilitate your whole-body recovery from exercise, but nothing more.

7.8. Forty Years of Bad Science?

One of the earliest formal investigations into the nature of nutritional ergogenics dates back to 1924. Researchers witnessed a dramatic improvement in marathon performance when participants were fed confectionary during the race. Compared to the previous year, marathon times improved and those that finished exhibited a lower incidence of hypoglycemia (low blood sugar). The study inspired a century of research into carbohydrate as a potent fuel for endurance sport. Today, sports drink manufacturers are keen to denote that their products are endorsed by science, but they're usually reluctant to provide the details. A comprehensive 2012 report[70] collated a series of claims relating to sports performance products. In one arm of the report, researchers asked that the manufacturers of a number of leading sports drinks provide supporting evidence for their product's effectiveness, but only one manufacturer – GlaxoSmithKline (GSK) – complied with the request. At the time of publication, this multinational conglomerate was the maker of Lucozade, a carbohydrate sports drink sold the following year to the Japanese company Suntory, for £1.35 billion. Distributed to countries around the world, Lucozade was marketed as an energy drink that could *Keep top athletes going 33% longer**. The ominous asterisk accompanying the slogan denoted an important caveat which will be shortly discussed. Over the years, Lucozade Sport has sponsored numerous major sporting events, individual athletes, and teams in the UK and Ireland, including the Amateur Rowing Association (ARA), FA Premier League, FA Cup, England Rugby Football Union, the England Football (soccer) Team, the Republic of Ireland Football Team, the London Marathon, Parkrun, Michael Owen, Steven Gerrard, Damien Duff, Ronan O'Gara, and Ben Wynne. From 2012, the McLaren Formula One team donned the Lucozade sponsor on the rear-wing. The bibliography supplied by GSK on the research relating to Lucozade was comprehensive, to say the least: 176 studies dating from 1971 to 2012. Of these, 106 were able to be critically reviewed. This number was culled to 101 clinical studies by excluding posters, theses, and unavailable articles. The company's compliance for data should be commended; indeed, most of their contemporaries didn't comply with the request, either due to a lack of available data, because the data didn't show positive findings, or the data were deemed to be of insufficient quality. However, what follows is the *British Medical Journal* (BMJ) summary of methodological problems that were unearthed on review, including some serious concerns on study validity. It's a critical lesson on the importance of study *quality over quantity*; collation of all the scientific research the world-over is inconsequential if it's not of sufficient rigor such that meaningful conclusions can be drawn.

Sample size. Statistical power is an important concept on which we touched briefly in *Chapter 4*. It's the likelihood that a study will detect an effect when there's an effect to be detected, and it should be discussed transparently in published research where high or low sample sizes might influence study outcomes. Few of the studies provided made even a passing allusion to statistical power, and only four of 106 provided a power calculation (an *a priori* estimate of the sample size required to ensure statistical power). Moreover, the average sample was just nine in number. This is problematic because small studies are systematically biased toward the intervention, and can overestimate the importance of results.

Outcomes that aren't externally valid. Numerous studies tested the sports drink against water, rather than the drink's nearest competitor. Moreover, they frequently assessed the effect of Lucozade on athletes who were partially carbohydrate depleted (following an overnight fast), not often a state in which one would commence a bout of high-intensity exercise. Finally, seventeen of the studies utilized exercise time to exhaustion as a means of performance evaluation. Not only are such exercise protocols rarely contested in real-world competitions, but they also show a high degree of variability (i.e., subjects find it difficult to perform consistently in such tests). Doubt must be cast over the ability of these studies to detect small changes.

Poorly designed studies. Most of the studies (76%) had a lack of allocation concealment and no blinding. Therefore, subjects were aware when they were supplied with sports drink. Given that the intervention was designed for the purpose of performance-enhancement, unblinded studies likely evoked a potent placebo effect, making it difficult to distinguish physiological effects from psychological ones.

Misleading claims. The most impactful claim, used broadly in the widespread marketing campaigns (and printed on the label), was that Lucozade kept *top athletes going 33% longer**. Specifically, when consumed during high-intensity exercise, the drink delayed fatigue and improved running capacity by 33% relative to water alone. The relevant study was conducted in just nine participants who'd fasted overnight (for at least 10 hours), performed a 15-minute warm-up, followed by 75-minutes of exercise, after which they were asked to run at a fixed intensity to exhaustion while drinking either carbohydrate or artificially sweetened water. While the outcomes, *per se*, aren't in question, the test conditions (overnight fasting, pre-fatiguing exercise) were contrived to benefit the carbohydrate intervention, and don't relate to that which might be contested in competition. This isn't high-quality science, and the procedure was repeated in a number of the studies discussed. Cogently, when subjects were fed breakfast before participating in exercise, as is usual in sporting competition, the effects of the carbohydrate drink were less pronounced[71]. The *BMJ* report also indicates that augmenting endurance performance by 33% is unrealistic in most models, analogous to quickening a marathon from three hours to two.

Lack of blinding. Only 36% of studies blinded their participants to the solution being trialed; i.e., 64% of tests were performed after participants received a sweet, orange glucose drink, or plain water. Given that participants were anticipating ergogenic effects of the carbohydrate solution, such an expectation bias almost certainly influenced the results.

To summarize, this is less an evaluation of the effectiveness of carbohydrate sports drinks and more a commentary on the disconnect between scientific claims and the validity of the data they invoke. GSK is a huge, multinational company with multiple sites and near limitless resources; yet, according to the *BMJ* report: *...if you apply evidence-based methods, 40 years of sports drink research does not seemingly add up to much, particularly when applying the results to the general public.* These data relate to a single sports drink manufacturer; the solitary one willing to comply with a request for evidence.

7.9. Supplements with Robust Supporting Evidence

Contrary to my somewhat negative evaluations, it's not all doom-and-gloom in the supplement industry. While true that there's a dearth of convincing evidence for many of these pills and potions, and that manufacturers are perpetually overstating the magnitude of any effects, there are a few supplements for which the evidence is more positive and on which there's general agreement on safety and efficacy. What follows is a brief overview of those supplements, although given that there are numerous published summaries devoted to each, I've spared you an exhaustive assessment of the decades of research.

Creatine. Although creatine is found in several foods (most abundantly in red meat and fish), it's unreasonable to obtain ergogenic concentrations without supplementation. Creatine is an organic compound stored in muscle cells that contributes to high-intensity, explosive movements. In fact, maximal exercise lasting less than ten seconds is fueled predominantly by the phosphocreatine system, of which free creatine is a crucial component. Supplementation increases the muscle's free creatine concentration, thereby facilitating recovery from maximal exertion. Creatine doesn't influence muscle growth or stimulate muscle protein synthesis *directly*; it facilitates *recovery* from short-duration, high-intensity bursts of activity, thereby increasing the volume of work that can be performed in a given session. Moreover, it's unlikely to benefit a single maximal effort. Nearly 30 years of research has failed to demonstrate any long-term negative implications of use, but some urge caution in this regard. Increased muscle concentrations of creatine lead to water retention and increases in body mass. It may not, therefore, be helpful for endurance runners or cyclists because it'll harm your economy.

Caffeine. This ubiquitous stimulant is found mostly in coffee, tea, chocolate, and many types of soda. Many years of research have established caffeine's role as an effective ergogenic for many sports. It exerts its effects via two mechanisms: first, by acting directly on receptors in the brain to increase cognitive function (e.g., reaction time) and mitigate tiredness; second, caffeine acts on muscle tissue to increase the strength of muscle contraction. It's unclear if your habitual intake (the amount you consume daily) influences the ergogenic action, but some studies suggest that reducing your daily intake will maximize caffeine sensitivity. Individual variance is pertinent; some respond sensitively to small doses, while others respond modestly to larger doses; thus, it's important to experiment tentatively to discern your individual limits. Finally, and critically, caffeine is dangerous if taken in large doses (it used to be banned in competition for this reason); a typical mug of coffee contains around 100 mg, and a single ergogenic dose shouldn't exceed 4–6 mg per kilogram of bodyweight (1.8–2.7 mg per pound).

Protein. Given the earlier overview, I won't indulge in further examination here. Needless to say, the research is unequivocal that protein is a crucial macronutrient for numerous bodily functions including muscle recovery and immune function. Exercising individuals (particularly athletes) require greater amounts than those who are sedentary and, although a food-first approach is generally preferred, some people choose to supplement with protein if they struggle to meet their daily requirements. There are clear evidence-based guidelines on daily amounts, doses, and timings and there's little excusing chronic overconsumption.

Carbohydrate drinks. Despite the summary surrounding literature on the Lucozade brand, a hundred years of research has established carbohydrate drinks as a key source of energy for long-duration endurance performance. It's critical in this respect to denote precisely when such drinks may confer a benefit. The body has a limited internal storage capacity for carbohydrate (around 500–800 g), whereas our storage potential for fat is almost limitless. During exercise, one can exhaust their carbohydrate reserves within a few hours (more rapidly at higher exercise intensities), at which time weakness, lethargy, nausea, and dramatic decreases in exercise performance will likely manifest. In contemporary culture, this phenomenon is what's known as *hitting the wall*.

Carbohydrate drinks (containing glucose, fructose, or a combination) provide additional fuel for the muscle cells, thereby sparing internal carbohydrate stores for the latter stages of performance. There isn't a great deal of convincing evidence to suggest that glucose/fructose drinks will benefit exercise lasting *less than* 90 minutes; however, for a marathon contested at a pace slower than the typical elite runner, or a triathlon/cycling race of equivalent duration, carbohydrate drinks have been consistently shown to improve power output and perceptions of effort. They can also be effective at kick-starting recovery if consumed within 1–2 hours of exercise. In almost every other scenario, such simple sugars have little-to-no function.

7.10. Supplements without Robust Supporting Evidence

Most of the others.

7.11. Food Intended for Sportspeople

The European Food Safety Authority (EFSA) is an EU agency that provides independent scientific advice on existing and emerging risks with respect to food products, or those marketed as such. The EFSA makes decisions relating to risk-management, and publishes reports and opinions on food and food supplement safety and effectiveness. The organization cooperates with the national food safety authorities of the 27 EU-member states, formerly including the Food Standards Agency (FSA) of the United Kingdom. In this respect, the EFSA is the European equivalent of the American FDA, and the FDA is authorized (under 21 C.F.R. § 20.89) to share non-public information with the EFSA regarding FDA-regulated products *as part of cooperative law enforcement or cooperative regulatory activities*. Despite criticism over the years for a perceived lack of vigor, both organizations authentically strive to regulate sports products. The wide array of published documents that are congruent with their efforts are for the most part freely available in the public domain.

In 2015, the EFSA published the report: *Scientific and technical assistance on food intended for sportspeople*[72], in which they compiled existing scientific advice in the area of nutrition and health claims. It's not the most accessible of documents, a protracted 32-page report littered with technical jargon. Nevertheless, there are some important take-home messages which I've summarized. First, they offer six statements denoting scientific points on which both the EFSA and the Scientific Committee on Food (SCF) agreed. These correspond, in the most, with that already discussed: (1) there is an essential role for carbohydrate in the recovery of normal function after strenuous exercise; (2) there is an essential role for carbohydrate, hydration, and electrolytes during endurance exercise, and for electrolytes in post-exercise rehydration; (3) protein has an essential role in the growth and maintenance of muscle mass; (4) micronutrients [vitamins and minerals] and long-chain polyunsaturated fatty acids have an essential role in body functions that influence health or performance; (5) caffeine has ergogenic [performance-enhancing] effects on endurance exercise; and (6) creatine has ergogenic effects in short-term, high-intensity, repeated exercise bouts. Beyond these consensus statements, a few additional comments denote that multivitamin supplements were likely unnecessary for everyone, and that doses of vitamin C above the Population Reference Intakes might be beneficial for athletes. Moreover, the EFSA didn't wholly agree on the long-term safety of creatine, particularly that above recommended values:

> The SCF considered that the information available regarding the safety of creatine supplementation was lacking or incomplete, particularly at doses exceeding the specifications above. Creatine ingestion prior to competition in the heat was discouraged, as it may interfere with water absorption and as there was no rationale for intake immediately before competition. The SCF found it unlikely that acute and long-term consumption of creatine supplements would impair kidney function in healthy individuals.

I discovered the EFSA report only after concluding this chapter, and then integrated its conclusions *post-hoc*. It was heartening, therefore, to yet again note different authorities arriving independently at the same general conclusions. This is an important consideration in the broader context because if data are reviewed objectively, in the entirety, using every skill to eliminate bias, then there's every likelihood that one set of conclusions will reinforce the other, time after time. And when new studies are added to the ever-growing reservoir of human knowledge – an integral and ongoing facet of the scientific process – then periodic re-evaluations of the whole will acknowledge the updates, and the conclusions will be revisited. Some contentions (e.g., *protein has an essential role in the growth and maintenance of muscle mass*) are established norms, and overthrowing such a tautology would require a great deal of very robust and compelling evidence. Critically, however, there's always an invitation to such a challenge. And that's science at its strongest; its greatest strength is its capacity for change.

8

TRAINING PROGRAMS AND PRODUCTS

gimmick.

A trick or device intended to attract attention, publicity, or trade.

8.1. Disproportionate Claims

The industry is characterized by a huge disparity between the claims made for product effectiveness (for any number of health or performance outcomes) and the evidence in support of them. A 2012 publication in the *British Medical Journal* – a respected leading general medical journal – studied and quantified this disparity[7]. The authors, largely from Oxford University in the UK, examined nearly 100 of the most popular magazines in the UK and the USA, and identified over 1,800 adverts for sports products, including footwear, clothing, electronic devices, and supplements. The adverts, from manufacturers as opposed to product suppliers, were included in the analysis only if they made specific claims relating to performance and/or recovery. Once compiled, the websites of those products were independently searched for any references that might support the claims; consequently, both the printed magazine advert *and* the corresponding website were searched for evidence. It was a thorough and comprehensive assessment. From the printed ads, 54 different products were identified for which there were 113 claims related to enhanced performance and/or recovery; an average of two claims per product. Nearly half of the products had been endorsed by athletes, but only one such product was advertised alongside any references that could be appraised by the researchers. The websites of 104 products were searched, revealing 431 claims, and thus an average of four claims per product, double that seen in the magazines. Online platforms were more conducive to the citing of evidence (i.e., downloadable PDFs, links to external resources, etc.); nevertheless, when asked for documentation in support of their advertising claims, several large manufacturers exhibited a striking lack of cooperation. From the *BMJ* study:

> We contacted 42 companies and received responses from 16, of which two were unwilling to share their research (Panache and New Balance), one provided a video of the product in use and said that this was 'sufficient' (Nike), one pointed to the work of one researcher but did not answer whether the company had any research on its actual product (Merrell), one responded that they would get back but did not, one declined due to staff absence and one directed us back to their website.

More than half of those websites making specific claims didn't provide *any* references, and many of those that were supplied were non-existent, duplicates, books without clinical studies, animal studies (including a 1930s paper on rat metabolism), conference abstracts without data, or online surveys for which data could not be obtained. Finally, of the remaining 74 studies that were offered as *evidence-for-efficacy*, none were systematic reviews, and 84% were judged to be at a high risk of bias because, for example, they weren't randomized studies.

Aside from a general over-reaching in the advertising claims, the analysis revealed further incongruities. First, there were scarcely any published systematic reviews or meta-analyses concerning the products on sale. This more often denotes insufficient evidence on which to perform such analyses, precluding a scientific consensus. Review articles are summaries of the research in a given area, and they contribute to the public understanding of a topic. As a career scientist, I've the training and insight to be able to determine, for myself, the strength of evidence in support of a product or a claim. My knowledge is also subject-specific, so when viewing an advert that makes a factual claim about an aspect of human physiology, I'm in a position to determine its truthfulness. But without subject-specific knowledge and training, a layperson may well depend on objective and comprehensive review articles to inform their decisions, and a dearth of reviews renders the public at a disadvantage when deciding on the relative benefit or harm. Second, where systematic reviews *were* available, they often went uncited. For example, carbohydrate sports drinks are among the most widely researched supplements in the world, with decades of research collated and summarized in countless narrative and systematic reviews on the history and evidence pertaining to their use. Yet, none of the adverts for carbohydrate sports drinks included any reference to such reviews. Why would this be? What's unlikely is that marketing teams are oblivious to the wealth of literature at their disposal, particularly given the historical evidence dating back to the 1920s. Moreover, they cannot be discouraged by any unfavorable consensus (as might be the case with other products), because despite the broad variability in study quality, the conclusions on sports drinks to benefit endurance exercise (greater than 90 minutes) are overwhelmingly positive. A more likely explanation is that scientific consensus is not deemed a valid means of influencing sales. Allow that last sentence to sink in; *the fact that systematic reviews are not cited as evidence for product efficacy suggests that scientific consensus is not deemed in marketing to be a valid means of influencing product sales.* There are more powerful tools at their disposal, including athlete/celebrity endorsements, and fallacious appeals to tradition and/or nature. Why submit evidence-for-efficacy, inviting discourse and scrutiny, when such evidence is not a prerequisite for sales?

The BMJ study concluded that *there is a striking lack of evidence to support the vast majority of sports-related products that make claims related to enhanced performance or recovery.* While this conclusion wasn't unexpected, it's the disregard for evidence that's a greater source of concern. Moreover, when studies are provided, they're so often of poor quality that it's problematic just arriving at meaningful conclusions; our solitary supposition is that *more high-quality research is needed.* There's also a chronic lack of agreement on what *evidence* should comprise. In place of objective, well-controlled studies that minimize bias, we're offered a lack of robust studies, academic posturing, and anecdote. Providing evidence-for-efficacy shouldn't be a symbolic gesture to placate the stuffy science elite; it should be routine, commonplace, and should be treasured, for it's the one system sparing us a sad regression to the scientific bedlam that characterized the middle-ages.

8.2. Recycling

Thus far, we've referenced sports science *products* in their broadest sense to encompass training programs, sports garments, supplements, diets, concepts, and philosophies. Anything that can be packaged and sold retains an inherent value and, therefore, has product potential. In *Chapter 4: Show Me the Research*, I summarized the practice of rebranding as it relates to sports products and programs, raising the idea that something *new* is not necessarily something *effective*, and such practices often comprise a rehash of old ideas. I'd like to explore this a little further with *exercise classes*, another type of product. Crossfit, for example, has taken the original tenets of Olympic weight-lifting, plyometrics, gymnastics, and strongman, and combined them into a novel and effective high-intensity training regimen. Independently, these modalities have been established for many decades, but are here integrated, rehashed, and sold with a new label and an image. With Crossfit, specifically, precepts of exercise and nutrition have been extended to a *training philosophy* and further into a *lifestyle*, to the extent that its proponents have become aggressively defensive of its critics, but no single modality is optimal for all outcomes. High-intensity aerobic training, to facilitate increases in cardiovascular fitness, can take the form of High-Intensity Interval Training (HIIT), Tabata, Fartlek, and Parlaaf, all premised on accumulating time above a certain heart rate or oxygen uptake, but with their own subtle nuances and ideas. The former of these, HIIT, has become especially popular and vogue in recent years, now incorporated into scientific studies which invoke the term HIIT in the title without evaluating any of its related protocols. Indeed, researchers the world-over are trading off the HIIT acronym in an effort to lend some validity to their work and increase citations. Some people train *only* in this exercise modality, insisting on its appropriateness for all domains of health and performance, for all ages and physical ailments.

There are at least 14 different types of yoga, and each can be performed in the heat or in the cold or in a thermoneutral environment. Moreover, while proponents of hot yoga claim that the conditions permit greater flexibility (a claim which isn't adequately supported by the literature), it's also claimed that excess sweating in the heat cleanses and detoxifies the body. Detoxification is also a purported benefit of cold yoga. Classes like Thai Bo and boxercise are now trending, as is the deliberate misspell of Combat to Kombat, affording it a novelty on the basis that it's edgy and Xtreme. Each exercise class comprises generic routines cobbled together from the detritus of others, rebranded as something innovative and new. Kettle-bell training is another regimen that has muscled its way into every gym in the country. This type of training can be extremely beneficial for certain outcomes, but some swear by kettle-bells for literally every physical ailment, and these assumptions are sometimes objectively wrong. The benefits of a novel training modality should always be evaluated, objectified, and contextualized. Data must be collected, and then theories formulated from the results; this is known as *deductive reasoning*. As Sherlock Holmes exclaimed: *it is a capital mistake to theorize before one has data; insensibly one starts to twist facts to suit theories instead of theories to suit facts*. Too often, we see instructors and trainers misappropriating training benefits to suit a pre-existing bias for a given modality.

The industry is saturated with an endless throng of training programs and products, scrambling for precedence over one another, all claiming to bestow the user with something original. They should all be considered only as tools; there's no one tool that's perfectly suited to every task, and you must pick and choose as appropriate to accomplish your goal. The goal itself should be clearly defined. Select the tool that works in that instance, and discard it when something else becomes necessary. And choosing the tool? Well, that requires an educated and informed insight as to the potential benefits and contraindications, careful deliberation as to whether the proposed benefits outweigh the risks, and a critical review of the evidence. The remainder of this chapter comprises short reviews on a series of sport- and health-related products chosen not because they're the most popular or evoke the most dichotomous views, but because they're among the most complex and compelling case-studies on which to bolster our skeptical toolkit.

8.3. Barefoot Running (Shoes)

People have been running barefoot for many decades, but for its current popularity, the practice owes a debt of gratitude to Christopher McDougall's 2009 book *Born to Run*. It's likely no coincidence that the Barefoot Runners Society was founded in the United States the same year the book was published; nevertheless, running completely barefoot is impractical, and not exactly conducive to life in the city, for example. The oxymoron *barefoot running shoes* was born to integrate modern contrivances with a deep-rooted yearning to uphold traditional values. Only since the 1970s have runners been exploiting technical advancements in footwear and donning their feet with a cushioned heel, arch support, and stiffened sole, whereas traditional barefoot shoes pertain to light, thin, or no-soled foot liners. Whether or not minimalist shoes accurately replicate barefoot running mechanics isn't clear. The difference between barefoot running and many other highly marketable commercial ventures within sport is that the major premise for its efficacy is quite plausible at first glance. From an evolutionary perspective, barefoot running is a natural extension of the habits of our hunter-gatherer ancestors who would walk, run, and hunt barefoot[73]. Proponents claim that it promotes a more natural running stride. Let's pause there for a second because their invocation of the word *natural* in this context should provoke a slight uneasiness in the reader. Whatever way you choose to define *natural*, there's no reason to think that it's necessarily good or better than *unnatural*; to assume so is to commit the *appeal to nature* fallacy. Moreover, there's much about the way that modern humans run (be it barefoot or shod in conventional trainers) that contrasts rather strongly with traditional practice. For example, running continuously on a concrete surface, as fast as possible (at or below the first lactate threshold) for 26.2 miles (42.2 km); and what about the sporadic twice- or thrice-weekly sessions, poorly paced on flat ground, that characterizes the training programs of many modern recreational runners? To assume that *natural* as our ancestors ran is compatible with contemporary culture is tenuous, at best.

In any case, there's a clear kinetic alteration with barefoot running when compared to shod running (i.e., running with modern running shoes). More specifically, when people run barefoot or with minimalist shoes, they alter their running gait to minimize the uncomfortable impact forces that'd have to be tolerated by heel-striking. The outcome is that barefoot running obligates a forefoot or flatfoot strike[74], which shortens stride length, and may reduce maximum impact forces through the ankle, knee, and hip[75]. Another positive consequence of the forefoot/flatfoot strike is that runners naturally implement a shorter stride length and greater stride frequency[76] with a more upright shank, all of which improve running economy[77]. Although the initial claims aren't unreasonable, supported by preliminary evidence, the long-term evaluation as determined by several systematic reviews suggests minimal (if any) benefit. The most recent studies in this domain bring into question the purported benefits of minimalist running shoes in mimicking habitual barefoot running[78]. Moreover, a 2017 review of 15 studies on 1,899 total participants observed no difference in relative injury rates between people who run habitually barefoot compared to those who run habitually shod, and no evidence for differences in motor performance[79]. Finally, a systematic review[80] of 20 papers that focused specifically on the transition from shod to minimalist shoes concluded that there was limited positive evidence of a transition effect on running economy (only four of 20 studies) and muscle development (only five of 20 studies): *the injury incidence comparing running during the [minimalist footwear] transition (17.9 injuries per 100 participants) to matched participants in conventional running shoes (13.4 injuries per 100) appears equivocal.*

Much of the research into barefoot running has been conducted on individuals who habitually run in normal trainers, but who were tested running barefoot in a laboratory environment. Their running mechanics, therefore, were unlikely representative of someone who'd chronically adapted to barefoot running. Furthermore, much is made about the barefoot condition being more economic relative to contemporary shoes, and this is misleading. Running economy denotes the volume of oxygen necessary to perform a given amount of work. In this context, it's not the shoe, therefore, that dictates economy but rather the running biomechanics. Runners frequently improve their economy by engaging a shorter stride length, which also happens to reduce maximum impact force through the lower-limbs; this can be achieved whether or not running barefoot. For instance, I've consciously shortened my stride length in recent years for ultra-marathon running; I land slightly more flat-footed than before, perceiving it to be more comfortable and tolerable over long distances. But there was no need for a change in trainer to influence my gait; it's a red-herring.

The case-reports published in favor of minimalist running have described the biomechanics of Kenyan distance runners, many of whom run habitually barefoot. While interesting, the anatomy/physiology of the study population differs from the average Westerner. The former have longer femurs (thigh bones), lower indices of both muscle and fat, smaller frames, and longer/more slender limbs. Consequently, Kenyan distance runners cannot be used as a universal reference for the efficacy of barefoot running. Finally, it's unreasonable to suggest that poor running mechanics (attributable to contemporary footwear) are exclusively to blame for injury rates in modern runners, because there are many instances of running-related, lower-limb injuries in barefoot runners, including tibial stress fractures. It may be that more robust (less injury prone) runners simply make better proponents of barefoot running, but this hypothesis remains to be tested.

It's unfortunate that the marketing rhetoric places such emphasis on *running like our ancient ancestors*. Recall the *appeal to antiquity*, which suggests that a practice is correct or better because it correlates with some past tradition.

Accordingly, it's a fallacy to assume that running barefoot is correct or better just because our ancestors did it; our ancestors also believed in demonic possession, and cannibalized each other, but we find these traditions less agreeable. As always, practices should be judged only on merit. If the weight of the evidence proves them beneficial, then they should be maintained and perpetuated. If not, then they should be discarded because sentiment is overrated.

To summarize, the premise of barefoot running is logical and the early (preliminary) evidence promising; however, the marketing campaign hinges on several logical fallacies and, more pertinently, the long-term evaluations from systematic reviews suggest little-to-no difference in injury risk and/or performance with minimalist versus contemporary shoes. Until convincing research is published to the contrary, we should be very cautious in actively promoting barefoot running as a replacement for modern trainers. A transition to the minimalist method isn't justified by the literature, but should one be attempted, a safe and injury-free transition may require several months of careful training, predicated by a robust periodized plan to gradually introduce barefoot shoes. Users will certainly need to engage in a strength and conditioning program to correct pre-existing strength imbalances that may be exposed by the transition. The reality is that most recreational runners will not have the technical knowledge, time, or inclination to engage in such robust prehabilitation, and therein lies a principal barrier to this alternative method. It seems that the footwear with which you're encumbered is less important than both *how* you run (your biomechanics) and the independent training performed to reduce the risk of injury.

8.4. Compression Garments and Tights

Shiny polyester and elastane are a frequent sight at endurance races, and although compression tights are the athlete's commemoration of a hard session, they were never intended for sportspeople. They were developed first to alleviate circulatory complications in patients who were to spend prolonged periods in a state of relative immobilization and, specifically, prevent deep-vein thrombosis (DVT). Blood circulates the body at a faster rate during ambulation or exercise to deliver oxygen and nutrients to the various muscles and organs. Muscle contractions literally squeeze blood through the vessels (a process known as the *muscle pump*), thereby facilitating the return of blood to the heart (venous return). But inactive patients – particularly those bedridden – exhibit less use of their muscles. They can suffer from peripheral blood pooling and, if susceptible, this can increase the risk of thrombosis (blood clotting). Compression socks were designed to provide graduated compression; that is, higher pressures at the ankle and lower pressures at the thigh, to squeeze the blood vessels, move blood along the limbs, mitigate peripheral blood pooling, and increase the return of blood to the heart. The optimum pressures for maximizing the flow velocity of blood through the veins were reported as 18–8 mmHg, at the ankle and thigh, respectively[81].

The successful implementation of compression tights in the clinical setting forged the path for a similar product (albeit it with go-faster stripes) aimed at exercisers and athletes. As is common, an established mechanism was extrapolated and co-opted for a novel purpose. Compression tights in sport have been widely studied, and the claims fall into two distinct categories: (1) that wearing compression tights *during* exercise can improve sports performance; and (2) that wearing compression tights *after* training can facilitate recovery by increasing blood flow via the same means as that proposed in patients. Despite the large number of studies, the waters are muddied by methodological differences and lack of agreement, as highlighted in a 2011 review[82]:

> The literature is fragmented due to great heterogeneity [diversity] among studies, with variability including the type, duration and intensity of exercise, the measures used as indicators of exercise or recovery performance/physiological function, training status of participants, when the garments were worn and for what duration, the type of garment/body area covered and the applied pressures.

With respect to recovery, there's evidence of local blood flow augmentation with graduated compression and this, in turn, would be expected to facilitate the removal of metabolic by-products. And in fact, muscle swelling and blood concentrations of creatine kinase (a marker of muscle damage) were lower when compression tights were worn following resistance exercise, relative to non-compression clothing[83]. These data seem to reflect general findings in the literature, but the effects are often small and inconsistent among investigations. A 2015 paper reviewed 23 prospective, randomized controlled studies, and found that *there are conflict[ing] results regarding the effects of wearing compression garments during exercise. There is a trend towards a beneficial effect of compression garments worn during recovery*[84]. A year later, another review summarized the indices of performance, and concluded that there were no statistically significant effects of compression garments on running performance times from 400 m to the marathon. They also concluded no effects of compression on maximal oxygen uptake (a measure of aerobic fitness), blood metabolites, or gas exchange kinetics. They did, however, observe *small* effects on exercise time-to-exhaustion, biomechanical variables, perceived exertion, and maximal voluntary contractions of the muscles[85].

A point of contention with respect to these data is the impracticality of blinding subjects to the intervention; i.e., one can compare the effects of compression to non-compression clothing, but subjects are always cognizant of the condition received. This could be why many of the benefits of compression manifest in activities easily be influenced by placebo (e.g., exercise time-to-exhaustion, perceptions of effort, perceptions of muscle soreness, short sprints, and time-trials[86]). Given that non-specific (placebo) and specific effects cannot easily be discerned, we must be conservative when interpreting the data.

Several specific claims require closer scrutiny. One manufacturer of compression clothing claims that their garments can *increase oxygen delivery to active muscles during motion* and *reduce lactic acid build-up to facilitate recovery*. I'm unfamiliar with any mechanism whereby compression tights might facilitate oxygen delivery to muscles, and there's no convincing evidence to support such a claim. The statement regarding lactic acid suggests overtly that reducing lactic acid accumulation might facilitate recovery. This stems from a pervasive misconception that lactic acid causes muscle fatigue when, in fact, studies dating back to the 1980s show that there's no such association. Besides, following the initial post-exercise period, rest alone will likely return blood lactate levels to baseline within the typical time between training sessions[87], and so using compression tights for this purpose alone seems redundant. Pertinently, many companies sell compression socks that cover only the ankle and calf (and sometimes just the foot) while citing research conducted on full lower-body compression. I contacted one such manufacturer to request evidence to support the assertion regarding their compression sock, but I was shepherded to a generic website exhibiting cherry-picked studies on lower-body compression tights that were scarcely representative of the literature. There's no validity to the extrapolation of data on lower-body or whole-body compression to a sock or anklet.

In summary, whole-body or full lower-body compression garments may expedite recovery by facilitating blood flow and reducing muscle inflammation when compared to non-compression clothing, especially if the hours following training/racing are relatively sedentary (i.e., due to travel or sleep). But there's no reason to think that compression is necessarily *better* than other means of post-exercise recovery. There's also poor legitimacy to the claims regarding performance and/or reduced risk of injury. Nevertheless, the use of compression tights during exercise for placebo effects, comfort, or warmth is likely to continue.

8.5. Altitude Training

The commercial altitude industry is underpinned by several decades of positive research, some in-vogue novelty, many unknowns, and a great deal of industry exaggeration. Despite several brief forays into the topic, it's now appropriate to explore the premise of altitude training and make a deeper dive into the literature. The potential for altitude to impact on exercise performance was exemplified at the 1968 Mexico Olympic Games where those athletes residing or training at altitude dramatically outperformed the un-acclimated lowlanders. These days, altitude training is common practice among the elite (the undulating plains of Tenerife with a peak elevation of 3,718 m [12,200 ft] have entertained many of the world's cycling elite including Bradley Wiggins and Chris Froome). Moreover, altitude camps have become a regular fixture in the programs of track athletes from the UK and the USA. Even athletes contesting explosive, high-intensity sports like mixed martial arts (MMA) conduct training camps at high altitude (e.g., New Mexico with a mean elevation of 1,700 m/5,700 ft). Chronically acclimatizing to altitude, or engaging in acute acclimation protocols, can be very effective at enhancing exercise performance, and may be crucial for mitigating the negative effects of competing at altitude. Nevertheless, the way altitude research is exploited by many facets of the health and fitness industry represents yet another unjustified extrapolation from findings in the academic literature. Altitude is a very complex area with many unknowns; accordingly, what follows is an abridged and simplified account of the current state of knowledge.

Endurance exercise performance is largely mediated by the body's ability to drive oxygen from the atmosphere, through the gas exchange surfaces of the lung, into the blood, and to the active muscles. At altitude, the partial pressure driving oxygen across the lung membrane and into the blood is reduced, thereby inducing a state of relative hypoxia (low O_2). Such an environment stimulates a cascade of physiological adaptations including elevated numbers of red blood cells that improve oxygen delivery[88].

A meta-analysis of 51 studies[89] discussed several of the most popular altitude strategies concluding that among the most effective for amateur and elite athletes was *live-high, train-low*, whereby athletes live for an extended duration at altitude, but train in normal conditions either by returning to lower elevations or by exercising with artificial gas mixtures. They also elucidated that the factors most likely to mediate performance were maximal aerobic power and/or placebo effects. Despite a general consensus that altitude training is a valid component of the athlete's toolkit, there are a number of complications – overlooked in commercial settings – that warrant consideration. First, many of the physiological adaptations evoked by an altitude sojourn diminish within a month of returning to sea-level; sporadic mid-season training camps may, therefore, be of limited utility. Second, prolonged stints at high elevations are not without their risks; immune suppression and disturbed sleep are common contraindications, with acute mountain sickness (AMS) probable in 20%–40% of people, causing potentially fatal complications to health. Third, the reduced oxygen partial pressure dramatically reduces the intensity at which one can train; accordingly, it may be counterproductive for sports that are dependent on high levels of muscular power. Finally, a recent narrative review[90] on the effects of altitude training on physiological responses concluded: *Despite the potential benefits arising from altitude training, its effectiveness in improving haematological variables is still debatable...* and this may be due to differences in the reported *hypoxic dose, training content, training background of athletes, and/or individual variability of EPO production.* In fact, altitude non-responders are common, and some individuals simply don't exhibit changes in red cell volume or aerobic capacity with such interventions[91].

These important caveats and confounders haven't contained the commercialization of altitude training, nor impinged on its fashionable status. Companies sell altitude training sessions as shortcuts to improved health and performance congruent with far-reaching claims, but sporadic and infrequent exposures are unlikely to evoke any cumulative effects. Other products like the so-called *elevation training masks* are incongruous because they provide resistance to breathing that purportedly conditions the respiratory muscles, but the oxygen content of air remains unaltered. Accordingly, the apparatus doesn't adequately replicate the conditions of altitude; the function belies the name. I contacted one manufacturer of elevation masks hoping for clarification of their mechanism, but received no response. While there are legitimate benefits to a carefully considered and properly administered altitude strategy, much of the commercial interest hinges on an extrapolation and exploitation of the basic science.

8.6. Electrical Stimulation Devices

Of all the sports products advertised through the years, abdominal muscle stimulators are among the first I encountered as a youth. Even as a child, active and health-obsessed, the *something for nothing* premise felt instinctively wrong. It's strange seeing these antiquated infomercials having endured the test of time, run and re-run *ad nauseam*. The fact that these devices remain in production suggests there's an active market. The premise is that electrical stimulation can be harnessed to evoke muscle contractions that build and tone, but without the effort and exertion we've come to associate with *getting in shape*. There are several stimulation techniques, widely studied in the literature. There's electrical *nerve* stimulation, a technique which constituted a large bulk of my doctoral research, which involves passing electric currents through the central and/or peripheral nervous system to induce involuntarily muscle contractions. Not only can this be used to cause *supramaximal* efforts during exercise or certain breathing maneuvers, the technique is widely studied in relation to spinal cord injury to stimulate intact neuromuscular systems and provide therapeutic exercise for muscles that otherwise wouldn't be activated. Another form of electrical nerve stimulation is used therapeutically to reduce pain signals that are fed back to the central nervous systems in conditions like arthritis. There's also transcranial stimulation which involves passing a small current through the skull to target specific brain regions that modulate certain behaviors. These are all legitimate fields of study, wholly different to the handheld devices you see in stores or on the TV. Ab belts directly stimulate the muscle into a modest contraction, bypassing the nerves. This important functional difference alters the proposed mechanism, so be wary of manufacturers attempting to misappropriate research that wasn't conducted on their devices.

The adverts palpably embody the *one quick fix* mentality that's at once the star and the scourge of the health and fitness industry. The infomercials feature models and bodybuilders with the implied premise that such a physique is obtainable by using the ab stimulator alone, and without the inconvenience of healthy-eating or exercise. Anecdotes also feature heavily in the marketing strategy, with claims of dramatic weight-loss and narrowed waist-lines manifesting after only a short intervention.

There is sparse research with ab stimulators, and that which has been conducted shows mixed results. For example, a 2002 study showed no changes in strength, body composition, or perceived physical appearance after using the stimulator thrice per week for eight weeks[92], results that were largely attributed to the poor quality of the apparatus. Two additional studies were supposedly conducted in 2002 – observing improved strength, endurance, and improved perceptions of body composition – but the studies are cited as *unpublished* reports, authored by the product manufacturer, and aren't available anywhere online in any form. A 2005 study on the same apparatus reports an increase in abdominal muscle strength and endurance in 24 adults who followed an eight-week intervention, but with no associated change in body composition[93]. Any legitimate scientist, on these data alone, should remain unconvinced; these studies need replication and the description of a clear mechanism-of-action. Despite the contradictory and equivocal findings, studies have mostly assessed efficacy in healthy but untrained people who, by definition, don't engage in regular physical activity. It's unknown, therefore, if electrical stimulation will benefit someone already fit and strong. A second contention is that in somebody overweight (or specifically, over-fat), the potency of the electric current, given its limited depth, will be largely mitigated in its passage through a dense layer of subcutaneous fatty tissue. The stimulator probably won't be effective in the population to whom it'll most likely appeal.

Electrical muscle stimulators are currently regulated by the US Food and Drug Administration. The FDA's threshold for evidence can be rather low, and for years it has dealt with criticism from government and non-government organizations. This includes a multi-million-dollar 2006 report from the Institute of Medicine, published in the *Lancet*[94], which highlighted major deficiencies in the current system. Nevertheless, I read cover-to-cover the FDAs report on electrical stimulators, and below have compiled the key points:

> Most electrical muscle stimulators (EMS) that have been reviewed by FDA are intended for use in physical therapy and rehabilitation under the direction of a healthcare professional… While an EMS device may be able to temporarily strengthen, tone or firm a muscle, no EMS devices have been cleared at this time for weight loss… or for obtaining

> "rock hard" abs... FDA has received reports of shocks, burns, bruising, skin irritation, and pain associated with the use of some of these devices. There have been a few reports of interference with implanted devices such as pacemakers and defibrillators. Some injuries require hospital treatment... Devices may only be marketed for uses that are established for the device or for uses that the firm can support with data. At this time, FDA is not aware of scientific information to support many of the promotional claims being made for numerous devices being widely promoted on television, infomercials, newspapers, and magazines.

Finally, despite the unconvincing data, the manufacturer's claims are still somehow dichotomous from the literature. The products are purported to: (1) increase muscle strength; (2) decrease body weight; (3) decrease body fat; (4) improve muscle firmness; and (5) improve muscle tone. Getting *six-pack abs* as advertised is likely the principal reason for investing in a device, but such an outcome won't manifest by increasing muscle strength alone. To expose the superficial abdomen, one needs to build and/or tone the muscle while simultaneously reducing the layer of subcutaneous body fat that obscures it. Reducing body fat requires burning calories (of fat), which doesn't occur without first increasing metabolic rate and/or reducing calorie intake. Artificially stimulating a small region of superficial muscles won't by itself facilitate a meaningful decrease in body fat. It's likely to be a fruitless hunt to use an abdominal stimulator without also implementing chronic lifestyle changes to develop muscles *and* reduce body fat.

8.7. Power Bracelets

The PowerBalance bracelet was first produced in 2007. According to the manufacturers, the bracelets contained *a hologram embedded with frequencies that react positively with your body's energy field to improve vitality, strength, balance, flexibility, and sports performance.* It sold modestly in its first year, generating a profit of just $8,000. While these trinkets might seem trite by contemporary standards, the company expanded at a dramatic rate and, in 2010, grossed ~$35 million. The bracelet was worn by myriad high-profile athletes including David Beckham, Paula Radcliffe, Roger Federer, Rubens Barrichello, Wasps RFC, and numerous actors and celebrities. And then suddenly, in November 2011, PowerBalance filed for bankruptcy following a damming report from the Australian Competition and Consumer Commission (ACCC) which deemed them to have engaged in deceptive marketing. The manufacturers issued the following statement:

> ...we admit that there is no credible scientific evidence that supports our claims and therefore *we engaged in misleading conduct.*

PowerBalance is the epitome of *pseudo*science in sport and the mass exploitation of scientific illiteracy. Despite its lack of empiricism or scientific credibility, the company was extraordinarily successful owing to its sophisticated marketing and fallacious advertising campaign. It's for these reasons that PowerBalance serves as an exemplar case-study with which to explore plausibility and evidence in the industry.

As previously decreed, let's first consider the plausibility of the product and associated claims. You'll already have identified the informal fallacy of an *appeal to authority*, whereby celebrities and famous athletes endorsed the product in lieu of credible scientific evidence. A reverence for these champion athletes and performers is often sufficient to deceive the credulous among us, in much the same way as the sponsor of your preferred sports team; it forges an association in your mind between the sponsor and your favorite athlete. The technical terms employed in the advert – *frequencies* and *energy fields* – sound superficially plausible to a non-scientist, but they don't reflect real-world phenomena. A second informal fallacy *blinding with science* has been invoked as a smoke-screen to distract us from the lack of a legitimate mechanism. The notion of a human *energy field* is erroneous. Humans certainly produce heat, fluid is constantly shifting from place to place, and human movement is predicated by electrical impulses that propagate the nervous system. But an energy field that might be harnessed by holographic technology; that's not science. In exercise science, the term *energy* has a very specific and conventional meaning; it's the capacity for a body to perform work, measured in joules. The New-Age usage of the term is more mysterious, ambiguous enough to be incorporated into any number of explanations in conjunction with other equally nondescript terms like chakras, auras, chi, and meridians, none of which have a cogent meaning in the physical world. Despite these definitional inconsistencies, how does one *imbed frequencies* in a hologram? And how would a hologram influence an *energy field*? And how could that possibly effect strength, vitality, libido, or anything else? These are questions for which there are no answers. The proposed mechanism is superficially scientific, but there's no prior plausibility to the product.

It's a decidedly poor beginning when a product exhibits no prior plausibility, congruent with a marketing campaign founded on a series of red-flags by invoking flawed logic and reasoning. But any misplaced reasons to believe the extravagant claims of the manufacturer rapidly diminish when the evidence-for-efficacy is scrutinized. During their brief but highly profitable existence, PowerBalance made no attempt to support their claims with scientific proof, and to this day, there's only a modicum of published studies that have assessed the product. There exist two explanations for this: first, PowerBalance didn't fund any external research into the product because they envisaged no positive results; and second, no research institutions were willing to squander resources to assess the effectiveness of a product with no plausibility. Unfortunately, empirical evidence is not a prerequisite to product sales, and the huge financial investments the company channeled into celebrity endorsements and viral marketing made the science relatively inconsequential. Positive reports were anecdotal only, with one website testimonial suggesting that users of the bracelet would experience *an extra boost in many areas of life including the office and the bedroom.* No wonder the bracelet was so popular!

In the years since PowerBalance, a number of similar sham bracelets have been manufactured, each with a subtle nuance to distinguish it from the one prior, but with the same distinct lack of scientific underpinning. One such product uses magnets, rather than holograms, to *alleviate pain and promote general wellbeing.* The so-called *magnetic bracelets* purport to improve circulation of the blood through the use of *two rows of bracelets for Super Magno Power.* According to the manufacturer:

> The magnetic effect comes down to haemoglobin, the iron-based protein inside red blood cells. In the same way that iron filings align themselves along the field lines around a bar magnet, so the red blood cells align themselves along the straight field lines of Tao and Huang's electromagnet.

It all sounds reasonable; red blood cells do contain hemoglobin and iron, and iron filings do align themselves along magnetic fields. But for this purpose, these basic biological principles have been falsely appointed. First, the magnets themselves are typically very weak, with scarcely the strength to penetrate the skin, never mind the capacity to affect deep body tissues and joints. Magnets do produce magnetic fields, but the bracelets use static magnets which aren't capable of influencing blood flow. Second, whereas metallic iron is ferromagnetic (has a strong response to an external magnetic field), the iron in red blood cells is weakly paramagnetic (has a very weak response to an external magnetic field), and the atoms in oxygenated blood are without the free electrons in their outer-shells that would render them susceptible to external magnetic fields. Rather fortunate, for otherwise a patient undergoing a standard MRI scan would likely explode, or at least bleed to death.

There's a littering of research conducted on magnetic bracelets, primarily in outcomes associated with pain-reduction, mostly of poor quality but with a few exceptions. For example, a 2009 study[95] assessed the effectiveness of both magnetic and copper bracelets on pain, stiffness, and physical function in 45 patients with osteoarthritis. Following a 16-week double-blind protocol, the authors concluded that magnetic and copper bracelets were generally ineffective for managing any of the assessed outcomes. In addition, they suggested that any reported therapeutic benefits were most likely attributable to placebo. Worthy of note, this study was published in the journal of *Complementary Therapies in Medicine,* commonly grouped with alternative medicine. It's a sobering moment of transparency when a journal devoted to alternative therapies reports a null finding with an alternative therapy. Still, critics maintain that the terms *complementary therapies* and *alternative medicines* are deceptive euphemisms designed to presuppose medical authority.

One seller of magnetic bracelets cited a slightly stronger 2004 study published in the esteemed *British Medical Journal*[96], which assessed whether the bracelets could reduce pain caused by osteoarthritis of the hip and knee. The study was a randomized controlled trial with three-arms: a strong magnetic bracelet, a weak magnetic bracelet, and a non-magnetic dummy bracelet. While the patients reported less pain while wearing the product, the authors were uncertain whether the response was due to placebo effects, especially given the poor validity of the self-reporting of blinding status. Accordingly, even those studies considered more robust are unable to discern effects that are independent of placebo.

In the skeptical community, there's a general contempt for unproven and/or *pseudo*scientific therapies. Some years ago, two brothers from Australia (now at www.scamstuff.com) conceived *The Placebo Band*; a bracelet similar in look and feel to the PowerBalance bracelet, with the same holograms, and with the same distinct lack of observable effect on the body. Much like its counterpart, *The Placebo Band* failed to improve health, vitality, or fitness, but then it was never designed for such a purpose. It's now sold for around $6.00, primarily as a statement of skeptical intent, with the word *placebo* inscribed in large white letters. Other skeptical outlets have since imitated the idea with the aim of expounding the exaggerated claims of the original product.

Despite the existent range of magic bracelets, only the PowerBalance brand has been appreciably exposed and reprimanded by regulatory commissions, and skeptics play a perpetual game of *whack-a-mole* with the range of products on sale. While the market is steadily diminishing, there are those still being deceived by products feigning real effects; for this reason, our efforts to confront and debunk must continue.

8.8. Nasal Strips

These white strips of plastic have for years been taped across the noses of countless soccer, football, and basketball stars. The plasters were originally designed to aid sleep by opening the nostrils to facilitate airflow. The website of one manufacturer claims that nasal strips are *clinically proven to relieve nasal congestion*, but I couldn't find any links to clinical research on their site, just plenty of customer reviews and anecdotes from those people experiencing better airflow during sleep. In fairness, the website makes no mention of exercise or performance effects, and such outcomes have likely been inferred by the sporting community who've been impressed by their favorite athletes using nasal strips at major competitions. There's only superficial prior plausibility to the effectiveness of this product; that is, the mechanisms by which nasal strips might improve exercise performance are intuitive but false. For example, it's sensible to think that having larger, more open nostrils might facilitate airflow and allow more air (and oxygen) into the lungs. But most healthy people exhibit adequate ventilatory capacity during exercise. If someone is tasked, under laboratory conditions, with breathing as deeply and as quickly as possible for 12 seconds, the total air displaced will easily surpass that displaced during maximal exercise. Accordingly, the machinery with which we inflate and deflate the lung is generally overbuilt for its purpose, and the drive to breath during vigorous physical activity (even at maximal effort) isn't enough to stimulate maximum ventilation. There are some instances in which breathing can be limited in healthy people, but oxygen delivery to the lungs isn't generally considered a limiting factor for maximal oxygen uptake. Also, consider that the large increase in ventilation that occurs during exercise is achieved predominantly via the mouth, and is unlikely to be influenced by narrowed nostrils.

The research is decidedly messy, with varied methodologies, populations, sample sizes, and outcome measures. Much of that published has studied the effect of nasal strips on nasal patency (openness) to relieve snoring and obstructive sleep apnea, but these data don't translate to exercise during which breathing rates are considerably higher. A 2014 review of literature[97] collated 17 studies on the effect of nasal strips on exercise, with only four studies reporting positive physiological or performance effects when compared to a placebo (i.e., with the strip positioned incorrectly). The other studies found no performance differences. One study found an improvement in localized effects (e.g., airflow and time-to-exhaustion when subjects were forced to breathe through the nose, only[98]), but again, these data aren't relevant during exercise when greater ventilations are achieved via the mouth. The literature consistently exhibits subjective effects, e.g., perceived exertion and perceived nasal patency. Accordingly, use nasal strips during exercise if you're satisfied to use a product that likely only influences your perceptions of airflow via the nose.

8.9. Respiratory Muscle Trainers

If the magic bracelet represents the straw house of the health and fitness industry, easily blown apart in a single breath from a skeptical wolf, then the inspiratory muscle trainer is surely the house built of brick. Very occasionally, products are developed the right way; a hypothesis is conceived, informed by years of subject-specific knowledge and research, rigorously tested by numerous independent labs, and the subsequent advertising claims based on findings from high-quality, randomized controlled trials. The marketing is underpinned by a rich empiricism. Respiratory muscle training, while not a panacea or a cure-all, is widely effective for a number of performance and clinical indications; more importantly, the claims are tempered and commensurate with the evidence.

The respiratory muscles are those that facilitate breathing, and they comprise the diaphragm, the superficial abdominals, and numerous accessory muscles that contract at different times and under different conditions as part of a wonderfully complex mechanical symphony that ventilate the lungs. Just like your legs would tire if you ran very fast for a prolonged period, your respiratory muscles can fatigue if you're breathing sufficiently hard for more than a few minutes, or if a clinical condition disproportionately increases their workload. There are consequences of such fatigue, which may include: (1) an inability to properly expand the lungs during exercise; (2) an increased breathing discomfort; and (3) a reflex which draws blood toward the respiratory muscles and away from those tasked with locomotion. Collectively, such responses can compromise exercise tolerance.

To continue the analogy, stronger leg muscles can exert more force than weaker ones. If one wished to improve their leg strength, they could engage in a conventional resistance-training program at the gym. Weight-training is inappropriate for the diaphragm which is located deep in the thoracic cavity; instead, it can be trained with repeated inward breaths through a semi-occluded mouthpiece that offers a resistance. The diaphragm and other inspiratory muscles must generate more force to overcome the resistance. As with conventional weight-lifting, the resistance is increased periodically and physiological adaptations render the muscles stronger and more fatigue-resistant.

The better (more well-controlled) studies tend to employ a *sham* condition, whereby the effects of normal inspiratory muscle training are compared to those induced by training against a resistance great enough to convince subjects they're doing something beneficial, but not great enough to elicit a substantial adaptation. Given that subjects are blinded as to which treatment they're receiving (real or sham), researchers can account for placebo effects. There exist reviews and meta-analyses in numerous areas including, but not limited to: the use of inspiratory muscle training to improve exercise performance[99,100], endurance performance at altitude[101], the management of asthma[102], the treatment of heart failure[103], and as part of a pulmonary rehabilitation program in patients with chronic obstructive pulmonary disease (COPD)[104]. The findings aren't overwhelmingly positive; there's a range in the extent to which people respond, and there's much we still need to elucidate regarding the precise mechanisms by which respiratory muscle training might influence exercise. There's also large variability in study design, making it difficult to compare among them, and there's a broad spectrum in the quality of available research (with low-quality, non-placebo-controlled studies at one end, and detailed, mechanistic studies at the other). We must, therefore, focus on the findings from the latter group. In general, training with the device appears to increase the maximal pressure-generating capacity of the inspiratory muscles which, in turn, provides a degree of protection from the consequences of fatigue. Training may also relieve some of the discomfort associated with heavy breathing during exercise. As a result, exercise tolerance may be improved.

The most widely available inspiratory muscle training device is sold by Powerbreath®. Their website makes one, multi-faceted claim: *Improves breathing strength and stamina, reducing fatigue.* The claim is specific, tempered, and in accordance with findings in the literature. For instance, there are studies supporting the notion that inspiratory muscle training can augment maximal respiratory muscle *strength*, others showing improved time-to-exhaustion, thereby improving *endurance/stamina*, and research showing objective reductions in the magnitude and prevalence of respiratory muscle fatigue (assessed using nerve stimulation). The website's opening tab isn't plastered with testimonials or anecdotal reports from athletes and patients who've benefited from the device; in fact, nowhere on the site are testimonials invoked. There's a link to a page of precautions and contraindications where consumers are warned, among other things, *not to make changes to any prescribed medication or prescribed treatment program without consulting your doctor.* At the foot of the homepage is a link to *Research Articles* which redirects users to the published and ongoing research, grouped by topic (clinical, performance, asthma, COPD, sedentary, etc.). Under the section dedicated to *Fitness and Sports Research*, there were six review articles cited. Accessing studies wasn't immediate; the site designers preferred to first offer a bitesize and accessible press-release on the relevant study and, in most cases, it required a few additional clicks to locate the original published manuscript. But all the cited studies are present and catalogued. That's not to say that quantity is preferable to quality; on the contrary, we must be wary of manufacturers that barrage the consumer with an overabundance of data, making it difficult to discern the good data from the bad. Moreover, the list isn't exhaustive or completely objective, and I couldn't locate a single study on the site documenting a non- or negative response, of which there are many in the published domain. Nevertheless, the in-house report on any given study appears to provide an objective interpretation of the findings. For example, one summary stated that: *A specific training of inspiratory muscles (Powerbreathe® Sports performance) increases the power of these muscles (voluntary and non-invasive tests).* The latter caveat is an important one, given that voluntary and non-invasive testing is considered less robust, and usually subject to greater bias. Despite the minor shortcomings, it's assuredly the best system I've seen from a commercial enterprise.

To summarize this chapter, advertising claims without adequate supporting proof are abundant, and for this, manufacturers face little consequence or accountability. Nevertheless, claims can also be reasonable, and the difference between the two is evidence-for-efficacy. I take no issue with products being sold on extravagant claims if those claims are supported by extravagant evidence. In an effort to compel the industry toward more competently supporting their claims, it may prove effective to consider the tenets of supply and demand. Consumers scarcely demand evidence for a claim, and so manufacturers are scarcely concerned with providing it; why should we expect any different? Selling products only on valid claims is restrictive, funding research is costly, and if the only products that were sold were those that'd been rigorously tested by independent researchers, we'd barely have enough to fill a single shelf in the sports store. It's our responsibility as consumers to insist upon more and better evidence, a demand we can enforce with our buying habits. Stop purchasing those products that you suspect are sold on false claims, and instead invest in those with proven efficacy. Manufacturers will soon be obliged to change their marketing strategies, and place greater emphasis on academic credibility. This won't necessarily result in the totality of products being designed in the same vein as the Inspiratory Muscle Trainer, i.e., with an authentic research question addressed with controlled studies, and marketed with claims commensurate with the evidence. And it won't preclude manufactures funding their own research (a conflict of interest) or, even worse, fabricating data to increase sales. But if we as consumers aren't compelled to demand better evidence, then we're surely part of the problem. Instead, let's be a part of the solution. Turn the notion of supply and demand on its head.

9

COMPLEMENTARY AND ALTERNATIVE THERAPIES IN SPORT

un·a·vail·ing.

Ineffectual; futile.

9.1. What is Complementary and Alternative Medicine (CAM)?

For millennia, many majestic animals like the African elephant and rhino have been hunted for their valuable ivory tusks. While sometimes whittled into ornaments and jewelry, ivory is also poached for use in ancient Chinese medicine; a branch of alternative therapy that's widely considered a *pseudo*science. China is the world's largest importer of smuggled elephant tusk, in part because it's believed that ivory can bestow great strength and virility, as well as treat a number of medical ailments including skin conditions, epilepsy, sores and boils, sore throats, anal fistulas, consumptive fevers, and osteoporosis. With such miraculous, almost supernatural claims, it's no surprise that the ivory demand is so profuse. The demand is only somewhat placated by the slaughter of around 30,000 elephants each year. Such prolonged and widespread poaching has provoked evolutionary pressure for smaller tusks. In other words, those animals with the largest and most impressive tusks have been hunted near extinction, often killed before propagating their genes, while those with smaller tusks have been spared. Consequently, calves are birthed with limited or no capacity for tusk growth, largely in the name of alternative medicine.

The terms alternative therapy and complementary and alternative medicine (CAM) are mostly synonymous. The pertinent difference between science-based practice and CAM is that the former prioritizes evidence above any-and-all conclusions that might result, whereas CAM is underpinned by pre-existing beliefs, bad science, and in many cases an anti-science dogma. Using tests performed under controlled conditions, alternative therapies have generally been found to have no effect beyond those that are non-specific (i.e., placebo). Some of the therapies discussed herein aren't considered CAMs, *per se*, but are included because they're alternative treatments for which there's weak supporting evidence. For the purpose of this chapter, I'll invoke the familiar abbreviation *CAM* to denote any alternative treatment, even though they may not be intended for clinical application. A founding premise for the forthcoming sections is that, contrary to some conspiracy theorists, the medical profession – funded largely by government bodies – is in the business of curing disease, not causing it. As such, if a remedy is proven to work above-and-beyond the non-specific effects of placebo, it's generally incorporated into mainstream medical practice. A treatment labeled *alternative*, therefore, by very definition hasn't surpassed the agreed pre-existing evidence threshold, and thus remains an inferior surrogate to more effective practices. The latter therapies thrive wherever industry doesn't govern itself to the same stringent standards as the medical profession. To invoke the words of Tim Minchin: *Do you know what they call alternative medicine that's been proved to work? Medicine!*

Much like any dealer in controversial or questionable practices, proponents of CAM usually fall into one of two camps. First are those who believe sincerely that the treatments and products they're selling have special properties that supersede conventional treatments. Such a group may use the products in their own treatment or practice, have experienced profound benefits, and want others to experience the same wellbeing, despite ignoring the science that's contrary to their beliefs. Often, believers are themselves victims of those individuals in the latter group, who are fully cognizant that dried human placenta and essence of lizard tongue are expensive *pseudo*-treatments without supporting evidence. Such individuals harness placebo effects to monetize CAM (it's a lucrative business, estimated to be worth $60 billion in 2018[105]) and exploit those of us less well-acquainted with the sciences. The camp into which a proponent falls is inconsequential if claims are judged objectively, on merit, with an empirical approach.

9.2. Why Do People Use CAM?

The reasons are multiple, and likely to differ depending on whether the therapy is purported to influence health or performance. A national survey[106] of ~700 individuals in the United States revealed that self-identified users of alternative remedies tended to be better educated, and reported a poorer health status than the general population. Nevertheless, it's unclear whether a poor perceived health renders this group more likely to try unproven treatments, or whether their poor health is a consequence thereof. The survey highlighted that the majority of alternative medicine users weren't dissatisfied with conventional medicine, but instead opted for alternative medicine because it more closely aligned with their pre-existing values, beliefs, and philosophical orientations toward health and life. To those of us who deal daily with science and *pseudo*-science, such a philosophical outlook is characteristic of CAM proponents, and it's usual for someone keen on homeopathy, for example, to also espouse the benefits of other natural and alternative remedies. It's analogous to the idea that a sportsperson who's taking one supplement has usually tried several; they're more amenable to, and accepting of, such practice.

Athletes may also use natural remedies and/or espouse anti-science rhetoric, but with different motivations and principles likely guiding their decisions. Athletes push, pull, twist, and tear at their bodies in an attempt to reach the lofty highs of optimum performance. Such arduous training and competing come at a cost; injuries, aches and pains are ubiquitous with life as an athlete. Competitive athletes are also perpetually seeking new and innovative ways to enhance their performance, with every 1% an eagerly sought-after commodity. The available data suggests that athletes may be among the highest users of CAM[107], although many stand by legitimate and evidence-based practices. The *pseudo*science of cupping – utilized by numerous athletes at the Rio Olympic Games, 2016 – is a prime example (see below), and interest in the practice grew eagerly following Michael Phelps' appearance in the Olympic final during which he sported a series of circular bruises across his back and shoulders. Boxer Floyd Mayweather has been pictured undergoing cryotherapy, which is the poorly evidenced practice of immersing oneself in air cooled to −200°C (−328°F). The NBA player Amar'e Stoudemire regularly bathes in red wine under the misapprehension that it'll expedite his recovery. Former professional soccer player Robin Van Persie, widely regarded as one of the best strikers of his generation, tried to facilitate his return from a ligament injury by massaging horse placenta onto his injured ankle: *I will fly to the Balkans to meet with a female doctor who helped Danko Lazovic [PSV Eindhoven midfielder]. She is vague about her methods but I know she massages you using fluid from a placenta. I am going to try. It cannot hurt and, if it helps, it helps. I have been in contact with Arsenal physiotherapists and they have let me do it.* The latter clause in his statement – that the procedure was endorsed by the team's physiotherapists – is indication enough that his support network was compliant in the practice, and this has been adequately scrutinized in our earlier discussion *The Price of Placebo*. In any case, CAM use by elite athletes is common, with their celebrity-status subsequently influencing use by the general public. Social media affords a platform for the instant dissemination of opinions, pictures, and stories, and is a principal reason why famous athletes – with their legions of followers – are pioneering population trends in the use of CAM[107]. A survey of over 300 intercollegiate athletes, from various sports, identified some interesting patterns in CAM use[108]. More than half had used CAM within the past year, and were significantly more likely to be female. The most common alternative treatments accessed were massage (38%), chiropractic (29%), and acupuncture (12%). Disconcertingly, 60% of athletes reported visiting their physicians regularly; this suggests either that discussions of CAM were being omitted from patient–doctor consultations, or that the physicians weren't successfully discouraging CAM use by their patients.

Many individuals use CAM due to their belief in the traditional philosophies associated with the treatment[109]. This explains why such practices prevail despite a lack of scientific evidence. Indeed, complementary and alternative therapies are pervasive, not because they're objectively effective, but often because they intersect with religious and cultural norms (e.g., ancient Chinese medicine). Not only might treatments be passed down from one generation to the next – granting them an authority on the basis of the *argument to tradition* – but criticism of such practice is poorly tolerated, and often considered culturally and/or ethnically insensitive. This can represent a major obstacle for those in search of truth via evidence-based practice. All ideas and systems of thought, be they scientific, political, religious, or cultural, must be subject to criticism and scrutiny. These systems are the lynchpins of contemporary society, and scrutiny is how we encourage good ideas and expose the bad, and there should be no exception.

Finally, due to its popularity and seamless integration into mainstream culture, many people are regular CAM users without ever realizing; massage and acupuncture are two common alternative treatments practiced by nearly all physiotherapists the world-over. When humans have a pre-existing propensity for belief in such practice, the instances in which they benefitted from the therapy (the hits) are remembered more often than those instances in which they didn't benefit (the misses). When conditioned to remember the hits and forget the misses, it only requires one positive experience for the individual to become a lifelong believer, despite a lack of demonstrable physiological effect. The take-home message is that CAM users already anticipate benefits, leading to the creation of myriad explanations to maintain the delusion. The lack of convincing evidence is of little consequence to true believers, because they have anecdotes reinforced by others with similar personal experiences. This is often sufficient to inoculate them against any amount of evidence and/or logical discourse to which they're exposed.

Even if void of *direct* negative effects on the body, there may be far-reaching consequences to the widespread use of CAM. As discussed in Chapter 5, condoning practices and treatments that we *know* aren't underpinned by scientific principles requires intellectual dishonesty. The most positive scenario is the inadvertent squandering of time or money, but the most negative sees CAM users forgo actual medicine in favor of natural remedies that don't actually remedy. While this may be of slight consequence for simple ailments or issues of sporting prowess, it isn't so slight for serious conditions that worsen, sometimes fatally, because they've been ineffectually treated.

Another issue, poorly considered, is that athletes often use a CAM supplement on the basis of its natural ingredients, neglecting that it may contain a substance that's banned in athletic competition. Complementary and alternative treatments aren't exclusively aimed at athletes, and despite being marketed on their *natural* components, they may contain some ingredients that contravene anti-doping prohibitions. While a supplement being found to contain an illicit substance would damage the reputation of a sports nutrition manufacturer (urging them to take precautions in the manufacture of their products), such isn't the case for manufacturers of oral CAMs. Never make assumptions on the constituents of supplements; whether they're harvested from a garden or a Petri dish, the safety of their ingredients cannot be guaranteed.

9.3. This is *Pseudo*science

Frequently, I've called upon this term, but nowhere is it more pertinent than in this chapter on CAM; it's time to give it a brief exploration. The word *pseudo* is derived from the Greek *pseudes* meaning false, and denotes a system or product with the appearance of one thing, but with the characteristics of quite another. *Pseudo*science, which is dictionary-defined as *a collection of beliefs or practices mistakenly regarded as being based on scientific method*, came into use in the late 1700s and, arguably, is now most commonly employed by skeptics and critical-thinkers to distinguish between real and fake science. The main distinction between scientific and *pseudo*scientific claims is that the latter aren't rejected on the basis of poor evidence. If we dealt solely in the realm of evidence-based practice, most CAM treatments would never have come to posterity; they'd have been cast aside like yesterday's broadsheet, or showcased in museums like medieval armor as a reminder of how far we've journeyed as the rational animal. But ideology is scarcely so easily discarded; it parasitically makes a host of tradition and superstition. In many instances, a lack of evidence only fuels the claims of the CAM proponent, who insists the benefits wrought aren't testable because they fall outside the realm of science, or are otherwise beyond our understanding. This is often borne of a general mistrust of the modern scientific process which espouses materialism and observation as grounding principles in the search for truth.

In their book *More harm than good; the moral maze of complementary and alternative medicine*, Ernst and Smith denote many of the problems manifest in CAM research, including: low participant numbers, lack of control groups, inappropriate control groups, use of surrogate end-points, misuse of statistics, *too good to be true* results, and fraudulent research. And while such flaws can be apparent in all research domains, collectively they seem characteristic of CAM. And where published research isn't available, scientific legitimacy is often feigned through the use of informal logical fallacies, and the retrofitting of modern principles to old practices.

9.4. Out With the Old, in With the New

Many of the CAMs featured in the remaining sections attempt to keep pace with the fast-moving empirical climate of modern science by replacing old and antiquated explanations with new and novel ones. Some CAMs (e.g., reiki, acupuncture, cupping) are hundreds or thousands of years old and were developed at a time in history when humans were primitive by modern standards or, at the very least, were largely ignorant of the laws of nature that underpin our present understanding. At one time, we conjectured that our planet was a Neolithic earth mother called Gaea who herself gave birth to the heavens. Later, dismissing these primeval notions as a fairy-tale, we became convinced the earth was flat, and that reaching its edge would risk falling into an eternal ether. Presently, some believe the flat-earth is contained by an enormous ice wall (Antarctica), beyond which no human can travel. In ancient Greece, we had no telescopes with which to inspect the celestial objects in the sky; thus, we were ignorant as to the significance of those sharp, distant twinkles. There was no germ-theory of disease; thus, we were naïve as to why people became afflicted with sickness. Early civilizations did their best with the philosophies at their disposal, but our initial attempts to explain natural phenomena were ill-informed and often incorrect; guesses that were largely dependent on our pious existence. Before the scientific revolution, we had to be creative with our explanations for natural phenomena. Fever was thus explained by demons and parasites, low mood by negative energy, and back pain attributed to blocked energy meridians in the body. These beliefs formed the basis of CAM. Leap forward several hundred years and The Enlightenment (post-scientific revolution) has afforded us new explanations based not on conjecture, but on observation and measurement, resulting in legitimate descriptions of nature. Yet for reasons aforementioned, some still cling to archaic ways.

It's exceptionally hard to escape the reality afforded us by modern science. Despite some obvious exceptions, explanations like blocked energy channels and demonic possession simply don't cut it in the modern world, and we're learning to cast aside these superstitions in favor of explanations that are congruent with our current understanding. As a result, the mystical claims of antiquity have been steadily replaced with a less dogmatic contemporary view that's more amenable to life in the 21st-Century, and all the contrivances available. Mystical beliefs have been replaced by science-sounding terms that *appear* to align with modern principles, but that actually fall short. Acupuncture, for example, was conceived as a means of unblocking body meridians and sustaining the flow of energy through the body, thereby curing the individual from a physical ailment. As understanding grew, and it became apparent that these mysterious explanations were wholly insufficient, proponents concocted new ones based on more modern principles; hence, dry needling was realized as a means of bridging an old practice with a new academic climate, despite the fact that the physical practice amounts to much the same thing. But science doesn't allow such concessions; one either follows scientific principles fully – prioritizing coherent logic and good evidence – or one doesn't. And yes, there's a great deal that remains unknown about the nature of existence and its overarching laws. But, like our distant ancestors before us, we must make the most educated and informed decisions permitted to us by our current systems of thought. The scientific process, when followed authentically, is the most sophisticated tool we have. We must enact good science, and let the axe fall where it may.

9.5. Cupping

One-by-one, the swimmers emerge from the athlete's tunnel to entrance music and rapturous applause. Behind the scenes, they've been finalizing their pre-race warm-up rituals to prepare their bodies a final time for the strenuous demands of high-level competition. Donned from head-to-toe in sponsored apparel, bearing the flag of their respective countries, the announcer blasts their names from the audio system. They strip jackets and hoodies and line up on the starting blocks. Some look relaxed and others more anxious, completely understandably, for this is the Olympic Games, the greatest showcase of sporting prowess on the planet. As they strike a statuesque pose, anticipating the starter pistol, one swimmer remains conspicuous, not just because he's the reigning Olympic champion, but because his otherwise clear skin has been tarnished by a number of large, ambiguous, purple and blue circular bruises across his back and shoulders.

There were several athletes competing at the Rio Olympic Games, 2016, who bore a similar pattern of blemishes resulting from an ancient Chinese therapy known as cupping. The procedure involves the placement of small glass cups onto sore muscles or sites of injury. Suction is created within the cup using a pump or the application of a heating mechanism. In the so-called *wet cupping*, the skin is then pierced with a lancet to provoke blood flow which, supposedly, removes stagnant blood and purges the body of toxins that are blamed for ailments including lower-back pain, muscle and/or joint pain, and fatigue, all of which are commonplace in the high-level athlete. Cupping was originally designed to stimulate *energy* flow along the body meridians. As discussed, these original notions have been largely replaced with more seemingly scientific ones so as to make the practice more amenable to modern culture. The characteristic bruising is caused by the clotting of broken capillaries.

It's opportune here to recall from *Chapter 4* that not all published studies are created equal. The extent to which we can make any inferences from scientific research is wholly dependent on the quality of the study and where it's published, not to mention the degree to which it represents consensus. With respect to cupping, most of the positive reports originate from China, the country whose cultures and traditions are heavily intertwined with the practice, and such an occurrence is indicative of systematic bias. A 2011 review of reviews[110] on cupping concluded: *the numbers of studies included in each SR [systematic review] were small... based on evidence from the currently available SRs, the effectiveness of cupping has been demonstrated only as a treatment for pain, and even for this indication doubts remain*. Such doubts persist because the positive findings are derived from poorly controlled, methodologically flawed, low-level studies, published in low-level journals or those dedicated exclusively to alternative therapies. The better (higher-quality) studies that control for placebo effects tend to show no positive results for any physiological or psychological outcome. Don't be deceived by the endorsements of several (very successful) elite athletes; in some instances, they are specifically targeted for their broad appeal and commercial influence.

Practices like cupping aren't exclusively benign. Skin burns are a common side-effect, with numerous published case-reports of infection and psoriasis. For example, a report from the Burns Registry of Australia and New Zealand noted 20 patients admitted to hospital, following cupping-related burns, during a seven-year period[111]. There are, of course, risks accompanying all medical procedures, and the physician is trained to make an informed risk-to-benefit assessment, in which the contraindications are carefully considered alongside the positives. Legitimate, evidence-based medical treatments are performed only when the risks are offset by the strong likelihood of positive outcomes. With cupping, specifically, there's no evidence of a positive outcome beyond placebo. For a competitive athlete, therefore, the potential loss of training time associated with such negative consequences could be costly, and the risks aren't justified.

As with all practices endured by athletes, rarely are they implemented following an isolated decision. There's barely an elite athlete in the Western world who isn't supported in their endeavors by a string of exercise professionals including medics, physiologists, and coaches. In the United Kingdom, the English Institute of Sport is the primary service-provider for the Olympic performance program. The UK system was modeled on the Australian Institute of Sport which operates on a similar model, although many sports also hire in-house scientists to work alongside those seconded. In the United States, athletes more often receive support from scientists employed by the national team which appoints sports physiologists, strength and conditioning coaches, psychologists, and medics to work alongside elite coaches in an effort to provide a holistic, high-performance training and competition environment. In Phelps' case, it's probable, therefore, that the therapy was endorsed and arranged by the scientific support network. Cupping found its way into his recovery regimen because it was justified by the ubiquitous power of placebo; i.e., scientific integrity was conceded in favor of the psychological edge expected to result from the therapy.

The further consequence is that celebrity endorsements of CAM tend to drive population trends. Following its appearance at the Games, cupping was featured in newspaper articles and TV specials, and a general fascination ensued as to whether cupping was responsible, at least in part, for Phelps' phenomenal success at the Games where he has won a cumulative 28 medals. While reporters were apprehensive to espouse cupping as a miracle therapy, rarely was it viewed through a scientific and/or evidence-based lens. With the media assuming a stance of such passivity, the door is left wide open for a more liberal and generous interpretation of cupping's effectiveness. All of this serves to propagate scientific misinformation. Elite athletes with their legions of fans are a prime target for the claims of *pseudo*scientific practices. This is a shame, especially for an institution like the Olympic Games which is celebrated for its ethos of excellence in practice.

9.6. Reiki

From the age of 8–21 years, I trained twice a week in the Japanese martial art of Karate-do, the way of the empty hand. My particular style, Shotokan, originated in the Japanese prefecture of Okinawa, which is a collection of islands in the south of Japan. In the mid-1400s, martial arts and the associated weapons were banned on the island by King Shō Shin, and the residents thus developed a system of self-defense, in secret, using farming tools and other commonplace implements. Karate itself was derived from these fighting techniques combined with Chinese martial arts, which the island inhabitants had acquired through trading routes with India. Some of the earliest forms of martial arts were developed by Shaolin Monks as a training regimen to harden their bodies and minds for the arduous task of serving their religion more fervently. Furthermore, most styles, save for the contemporary sport-based incarnations like MMA, originated in the East and are influenced heavily by Eastern philosophies and spiritual teachings, particularly those of Hinduism and Buddhism.

During my time as a diligent karateka (karate student), we were encouraged to study chi (or *qi*), a sort of universal life force. We'd often speak of energy and energy flow pragmatically, as if these mystical powers were tangible qualities that could be harnessed. A principal teaching, not unique to our particular style, was that the power in a strike originated from the *Hara*, a Japanese word which translates literally as *lower-abdomen*, but harbors a much broader and richer implication. The *Hara* encompasses a place in the body where the physical, psychological, and spiritual dimensions are unified, and it's claimed to be the contact center between the body and the soul. To detach oneself from the notion of such a spiritual force would have you labeled a naysayer and a bad influence on your peers. In fact, my old sensei (teacher) believed that he could summon energy from the *Hara*, and transfer it through the palms of the hands to encourage emotional or physical healing in a patient. Mr. Miyagi, legendary sensei from the Karate Kid movies, used the same energy to heal Daniel LaRusso and his dilapidated limbs after a fight. This notion of *healing hands* is the premise of the Japanese alternative therapy, reiki. For many, there's a keen fascination with the notion of finding spirituality in a self-evidently physical world. It brings a degree of comfort to acknowledge that there may be unknown entities offering a glimpse of realities beyond our own, which itself appears bland and pedestrian by comparison. Such a desire for the legitimacy of ethereal magic is largely responsible for reiki.

Before a statement on the evidence, let's consider the plausibility of reiki and/or therapeutic touch. The premise of the therapy is a *healing energy*, the existence of which nobody has ever been able to positively measure, quantify, or prove; this is a bad start. Given that reiki is a metaphysical assertion, it's a damming indictment of the state of modern scientific research that anyone was able to justify study into the technique in the first instance. As is characteristic, the research into the treatment comprises predominantly low-quality studies. Indeed, when a 2015 Cochrane review[112] applied stringent inclusion criteria such as randomization and rigorous methods to assess the evidence for reiki as a treatment for anxiety and depression, only three studies of an acceptable quality were found. Many studies pit reiki against control groups that don't receive any intervention; as such, the subjects aren't blinded to which treatment they're receiving, and placebo effects ensure to explain any positive outcomes. This is as true for *any* CAM treatment as it is for reiki. By way of example, consider a 2014 review of reiki[113] collating 12 articles which met their inclusion criteria. They concluded that *there is evidence to suggest that reiki therapy may be effective for pain and anxiety*. Yet, a closer scrutiny of that inclusion criteria revealed a decidedly relaxed approach in which numerous studies without sham treatments were admitted. A truly valid study in this respect would compare reiki responses to those evoked by an actor performing the same treatment sequence as the reiki practitioner. The *only* study in the review to employ such a sham treatment[114] independently concluded:

> ...Reiki is no more effective than mimic-Reiki in decreasing pain perception and improving walking distance in subjects with PDN [painful diabetic neuropathy]. However, the reduction of pain symptoms observed in both treatment groups is consistent with the concept that the formation of a "sustained partnership" between the health care provider and the patient can have direct therapeutic benefits.

Another well-controlled study providing a rigorous, authentic test found that reiki and sham-reiki interventions *both* statistically improved patient comfort and wellbeing, but with no discernible difference between them[115], further supporting the notion that positive patient outcomes are attributable to the time and attention patients receive from the practitioner, rather than any physiological benefit of the treatment.

A frequent occurrence in the world of CAM is the willingness of proponents to consider *real* and *placebo* effects as equivalent, but it's a false equivalency. Most such treatments are characterized by a pattern of non-specific placebo effects, meaning that they are psychosomatic. This occurrence is so frequent that CAM proponents are in the midst of a campaign to rebrand CAM as exhibiting efficacy *through the power of placebo*. Imagine hiring a coach for your upcoming 5-km track race. You pay good money in exchange for months of personal advice, session monitoring, and one-to-one training, only for the coach to admit the day before the race that their entire program was premised exclusively on a supposed *psychological* edge, and that months of work had failed to induce any physical changes. Irrespective of the race outcome, would you feel that you'd been dealt with honestly? Would you consider your investment justified? Would you recommend the coach to fellow athletes, close friends, and family? I'm supposing that you'd answer with a resounding *no* to all. You'd paid for one service and been given another, and would likely regard the practice as disingenuous. Yet, selling real, physical outcomes that work only in the context of placebo amounts to the same thing. If you're an exerciser, an amateur or professional athlete, why would you potentially squander your precious time and money on a treatment built on such shaky foundations? Steven Novella describes this co-mingling of placebo with an ineffective treatment as the '*Part of a Complete Breakfast' Fallacy*:

> Even as a child I recognized that when a commercial advertised their pastries as being part of the complete nutrition offered by an otherwise nutritious breakfast, the pastries were nutritionally irrelevant. They added nothing, and the commercial was being deceptive in trying to make me think that they were nutritious simply by their proximity to a nutritious breakfast.

While practices like acupuncture and cupping make an attempt at scientific coherence with the supposition of broad physical mechanisms, reiki is firmly premised on metaphysical notions of spiritual energy.

9.7. Acupuncture

Acupuncture is a fine example of how a procedure, originally underpinned by energy and other mysterious forces, can feign legitimacy to find its way into mainstream use. It's a vestigial practice from traditional Chinese medicine which has found a home in physiotherapy clinics around the world and, as a result, is widely accepted by the general public as a valid therapy. Nevertheless, the bulk of the high-quality evidence simply doesn't support acupuncture as an effective treatment. In its infancy, spearing small needles into the skin at acupuncture *sites* was purported to stimulate energy flow along body meridians. Such meridians would become periodically blocked and cause any number of physical ailments including irritable bowel syndrome, carpel tunnel syndrome, lower-back pain, osteoarthritis, chronic pelvic pain, and numerous others. Dry needling is the 21st-Century's attempt to replace 2,000 years of superstition with a degree of scientific credibility; it utilizes different needles and professes a different mechanism, but there's a close agreement between the two practices. The evidence for both acupuncture and dry needling comprises an abundance of data of varying quality, methodologies, and outcomes, requiring a certain degree of skill and patience to discern the good studies from the bad.

One barrier to the interpretation of data is that there's no distinct mechanism of action. Dismissing the notions of energy meridians and chi, practitioners argue that placing needles into *trigger-points* provokes the release of natural pain-killers (e.g., opioids) into the blood. Numerous other mechanisms have been proposed, including Pavlovian conditioned reflexes, nerve segment theory, gate theory, somato-autonomic nerve reflexes, stimulation of mechanoreceptors, and increased muscle blood flow and oxygenation. But these are *proposed* mechanisms, none of which have been clinically demonstrated, and there's a distinct lack of information on the associated anatomical structures. Some excellent summaries on the wealth of data have been compiled and collated by Science-Based Medicine (www.sciencebasedmedicine.org) and the NHS (www.nhs.uk). Depending on the review or meta-analysis you read, the conclusions on the efficacy of acupuncture span a broad spectrum; from no effect[116], to unconvincing[117,118,119,120] to slightly positive[121]. Several such reviews bolster their reference count by relaxing their study inclusion criteria, admitting studies of various standards (i.e., those with both rigorous and flawed methodologies). The weaker experimental designs which report positive outcomes, therefore, skew the conclusions of review articles, leading to a general overestimation of the benefits.

In instances where studies do show statistically significant effects of acupuncture, on pain-relief for example, the effects aren't considered to be *clinically* significant. Although this important distinction was discussed earlier in a less curtailed fashion, a result can be *statistically* significant because the difference between outcomes passes the threshold of an arbitrary p value (usually set at 0.05), but something *clinically* meaningful must have practical importance. Controlled studies show that acupuncture has no practical importance. In light of the unconvincing evidence, the necessary time and expense of implementation, and the risks associated with acupuncture, many clinicians feel that to conduct further research would be to squander valuable resources, as well as patient and researcher time.

Despite the conclusions afforded to us by the bulk of the available review studies and meta-analyses, acupuncture as a therapy is pervasive. Most of the research has been conducted in humans, but many veterinary surgeons appointed in animal rehabilitation centers administer *doggy acupuncture* to canines who, for one reason or another, are exhibiting pain. If we're to acknowledge the conclusions from human studies – that acupuncture works primarily via non-specific (placebo) effects – then it's not clear how the treatment can be justified on an animal with no concept or appreciation for the nuances of the procedure. The effectiveness of animal acupuncture is dependent on a clear and very precise mechanism that pertains to a physiological response, and there's no evidence that such a mechanism exists. Indeed, a systematic review of studies on the clinical evidence for/against the effectiveness of acupuncture in veterinary medicine was published in the *Journal of Veterinary Internal Medicine*[122]. The review comprised 14 randomized, controlled trials and 17 non-randomized trials, concluding that *there is no compelling evidence to recommend or reject acupuncture for any condition in domestic animals*, although the authors did suggest that there was some encouraging data that warranted further study. But to assume a generous stance, the latter statement doesn't justify the expensive and widespread use of veterinary acupuncture, particularly not when there are numerous other more effective, more affordable, treatments available. Ours is a climate in which *encouraging data* is apparently sufficient basis for a lucrative industry.

My concluding remarks on acupuncture are to highlight the very divisive nature of the therapy and its strong ideological underpinning. At a sports medicine conference I recently attended, I became embroiled in an aggressive discourse with an exercise physiology doctoral researcher who'd recently attended a workshop on dry needling. One of the tenets of a productive and logical discourse is that one should aim to be as generous as possible, as well as charitable and accepting, of the opposing arguments. In doing so, you increase the likelihood of arriving at an authentic and valid conclusion that wasn't the result of your pre-existing bias. The most generous we can be with respect to acupuncture is that there's no good evidence-for-efficacy. The proponent conceded that while she'd *not actually looked at any of the published data on the efficacy of dry needling, she 'knew it had worked' for her in the past.* This was a Ph.D. student, a personal trainer of over a decade, an individual with a degree and master's in the exercise sciences, attending an international sports medicine conference, founding an argument for efficacy on the basis of *it worked for me*. It's an exemplar of how an educated scientist can arrive at such a self-evidently uneducated conclusion. As critical-thinkers, we must endeavor to transcend ideological stances and strive for a greater understanding of the myriad ways humans can delude themselves through inherent biases and the placebo effect. Not to suppose that we're all deluded all of the time, only to acknowledge that we're fallible and to make allowances for it. That's why systematic, controlled, and objective scientific research is so crucial because it's designed to account for such bias and ideology. To take the pro-acupuncture position because of a positive experience, without having read or assimilated any of the overwhelmingly unimpressive research, is analogous to purchasing a shiny new car because sitting behind the wheel made you *feel* good, despite it having no engine or transmission.

9.8. Traumeel

For many years, non-steroidal anti-inflammatory drugs (NSAIDs), like Ibuprofen and Asprin, have been a mainstay for athletes in treating injury and reducing inflammation. But NSAIDs can have serious adverse effects on cardiovascular, musculoskeletal, gastrointestinal, and renal systems; the scientific recommendations, therefore, are to avoid NSAIDs during exercise, particularly ultra-endurance activities during which the renal system is taxed more substantially. The recommendation to abstain from NSAIDs during exercise has been exploited by the alternative therapy movement in the production of Traumeel, which was marketed specifically to treat sports injuries, inflammation, and pain. The laundry list of natural ingredients includes extremely dilute concentrations of: *Aconitum napellus, Matricaria recutita, Arnica (0.0015%),* and *Echinacea.* As is the scientific precedent, the data has its own voice. There are several studies suggesting that Traumeel might be a reasonable alternative to NSAIDs in facilitating recovery from repeated bouts of exercise[123] and in the treatment of musculoskeletal injuries[124]. Moreover, a double-blind, randomized, controlled trial on the effects of Traumeel (Tr14) versus placebo in 40 healthy adults concluded: *Tr14 might promote differentiated effects on the exercise-induced immune response by (a) decreasing the inflammatory response of the innate immune system; and (b) augmenting the pro-inflammatory cytokine response*[125]. Despite the encouraging early data, there are no reviews on the topic to establish a consensus, and further studies are needed before Traumeel can be confidently prescribed for injury management. Moreover, that which has been published suggests a meaningful benefit, but the observations must be put in the context of other anti-inflammatory treatments purporting to confer similar outcomes. Worthy of consideration is that Traumeel is marketed as a homeopathic remedy, although discussions are ongoing as to whether it can be classified as such, since traditional homeopathic products don't contain active ingredients. In 2017, the UK National Health Service ceased funding prescriptions for homeopathic products on the conclusion it was *at best a placebo, and a misuse of scarce NHS funds.* Accordingly, Traumeel manufacturers might do well to distance themselves from discredited practices.

9.9. Yoga

Yoga was established in Northern India around 5,000 years ago. Although originally developed as an aid to meditation, religious and spiritual practice, the Western incarnations comprise numerous poses, often stretches called *asanas*, almost at the exclusion of other activities. Yoga retains a heavy emphasis on spiritual attainment but, for many, it's a physical activity practiced for strength and flexibility. Many people also use yoga as a means to long-term weight-management, integrated into an active lifestyle and healthy eating regimen. A 2016 review[126] determined that, *despite methodological drawbacks, yoga can be preliminarily considered a safe and effective intervention to reduce body mass index in overweight or obese individuals.* Nevertheless, although yoga appears to be an evidence-based alternative for reaping the benefits of regular physical activity, it's just that – an alternative. There's nothing exceptional about yoga; it shouldn't take precedence over other forms of exercise, and it has no unique healing properties in and of itself. This is contrary, however, to what is claimed by many yoga instructors (or *yogi*) and it's this overreaching that contravenes the rules of fair-play. Yoga is, therefore, a prime example of the co-mingling of reasonable and unreasonable claims.

I've attended many yoga classes over the years, from conventional Hatha yoga (involving postures, stretching, and breathing exercises) to Bikram yoga (similar exercises performed in a hot room). My reasons for attending were conventional, but rooted primarily in the physical, i.e., to develop core strength in support of an injured lower-back (although due consideration needed to be given to some of the stretches requested of me), and improve lower-limb mobility. In fact, I've been afforded many of the benefits claimed by contemporary yoga practitioners, but too frequently has *pseudo*science crept into mainstream practice. A venture wayward of mainstream was indicated when, on entering the place, I found on their bookshelf volume after volume on *Foods for Healing*, and *Energy Flow*. During the session, while clutching knee to chest in a position to stretch the lower-back and gluteal muscles, the instructor informed the class that such a stretch would boost the immune system and improve liver function. Immune-boosting, liver-strengthening stretches are where I draw the line.

A common finding with many yoga schools (indeed, also with most forms of meditation) is that the Western practice is ubiquitous with the traditional Eastern versions that were conceived with religious, spiritual, and supernatural overtones. In fact, it's a near impossible task to find a yoga or meditation class that makes a clear distinction between the real physical benefits of the exercise and the imagined spiritual ones, although some such classes do exist. While for some, the spiritual entanglement with the physical is a keen attraction, for others (myself included) the many genuine physical benefits (like improved flexibility, stability, and core strength) become somewhat lost amidst a string of unreasonable claims, and the latter simply undermine the former. I've no doubt that many practitioners dismiss these outlandish claims as harmless cultural relics, but an *immune stretch* is demonstrably false. It seems a shame to undermine the legitimate benefits of yoga by conflating them with the illegitimate.

9.10. Cryotherapy

As a potential therapeutic tool for athletes and exercisers, cryotherapy ticks a lot of boxes; it's easy to administer, it's accessible (if one already owns the kit), it's quick (just a few minutes, per session), and – above all – it's theatrical. These reasons alone seem sufficient for the inclusion of cryotherapy in the rehabilitation/recovery programs of many professional sports clubs and performance institutes.

The practice requires the individual to enter a tank in which the air has been cooled, using argon and liquid nitrogen, to a temperature range of −150°C to −200°C (−238°F to −328°F). Four minutes of exposure is sufficient for the therapy to exert its purported effects. The physiological premise, taken from one of many cryotherapy marketing centers, is that cold temperatures on the body's periphery provoke a vasoconstriction (narrowing) of blood vessels which, in turn, reduces blood flow to inflamed muscles and joints. This notion of utilizing cold temperatures to mitigate inflammation is an oft-cited mechanism (e.g., a cold compress on a site of injury). While it's reasonable, care must be taken not to extrapolate the mechanism beyond that which can be supported by the evidence. There's more, however, to the suggested mechanism, and this is where the claims become emboldened and exaggerated. It's purported that while blood flow to the periphery is restricted due to vasoconstriction, blood is retained in the body's core where it becomes enriched with oxygen, enzymes, and various nutrients. When the subject vacates the tank and temperatures re-stabilize, blood then floods the muscles providing a *rush of endorphins* evoking a series of positive sensations. One website continues on how cryotherapy improves muscle recovery by *decreasing cell growth, increasing cellular survival*, and *boosting the player's immunity*.

The opening statement about vasoconstriction and reduced muscle inflammation is a valid, albeit contested, mechanism discussed at length in the scientific literature. This isn't to say it's necessarily valid in this context. The more sensational claims about immunity and cellular survival are what sprung the red-flags. There's also some confusion about the therapy *decreasing* blood flow to the muscles to reduce inflammation, but subsequently *increasing* blood flow to the muscles to deliver oxygen and vital nutrients; so, which is it? The description of blood becoming *enriched with oxygen and nutrients at the core* is erroneous; except in certain pathological conditions or in hypoxia, the systemic blood is always enriched with oxygen and nutrients. Another red-flag is the featured caveat: *These statements have not been evaluated by the FDA. These products are not intended to diagnose, treat, cure, or prevent any disease.* On reading such a statement – be it on a website, on the label of a supplement, or at the foot of the screen during a TV commercial – take a moment to consider its meaning; the claims being made have not been formally tested, and the therapy *isn't intended* to treat any diagnosable physical or emotional ailment. Contrast this with a drug developed for clinical use, which has been *specifically designed and manufactured* for the purpose of treating a physical or emotional ailment. The caveat is present when the product hasn't surpassed a minimum evidence threshold, and to permit the manufacturer an escape in the instance that somebody tries to use the procedure to treat something important.

Research with competitive athletes is scarce, and the widespread use of cryotherapy in professional sports is, therefore, slightly impetuous. The few existing studies tend to report on perceptual responses, e.g., data collected via questionnaires or numbered scales pertaining to feelings of muscle soreness. Perceptions are corruptible, easily influenced by any number of factors (by definition, they're not objective). And while perceptions are important for elucidating subjective experience, they serve only to interfere in matters of objective truth and establishing whether or not there's a real physical effect of an intervention. The placebo effect can also, at least in part, explain some of the differences seen in muscle *performance.* Consider the following study design: a group of 12 exercisers perform some fatigue-inducing protocol which is also designed to cause muscle soreness. Perhaps they perform some repeated maximal sprints, or multiple sets of heavy resistance exercise, during which their baseline performance is monitored. The group is then exposed to a cryotherapy intervention, after which the exercises are repeated in order to detect the magnitude of the functional decline. On another occasion, a week later, the group performs an identical protocol but in place of cryotherapy they're made to sit, passively. Muscle performance after either intervention will decline, regardless. However, one might expect that a real effect of cryotherapy would manifest as less of a decline in muscle function, relative to sitting. However, in such a scenario, the exercisers are likely to anticipate an improved recovery following cryotherapy; for one, they've likely heard of its beneficial effects and they're cognizant of the study purpose, but they're also being exposed to an interesting, novel experience. Their *expectation* of a difference might be all that's required to facilitate improved performance in the subsequent tests, especially when compared to a no-intervention control. This type of study design is fundamentally flawed, and would easily skew the data in favor of a cryotherapy intervention. In the research, it's this type of study that's more commonly published. This scenario exemplifies the importance of conducting sham-controlled studies, so that researchers can discount contamination by placebo effects. After all, facilitating recovery when compared to passive sitting isn't a profound observation, but doing so relative to other recovery interventions has far more meaningful implications, and these are the comparisons that should be made. As is acknowledged by the stronger review articles, the difficulty of *blinding* subjects to the recovery protocol makes it nearly impossible to exclude placebo effects. However, I'm of the opinion that a sham-cryotherapy trial could have been performed at temperatures less likely to exert positive effects. Indeed, if the therapy is truly underpinned by a physiological mechanism, you'd expect there to be a temperature-dependent response, i.e., a small effect at conservative temperatures, with more pronounced effects at extreme temperatures.

Certainly, more research is needed, not only with sham-interventions, but with bigger sample sizes. A Cochrane review[127] (Cochrane is a globally respected, independent network of scientists) assessed the efficacy of cryotherapy in treating muscle soreness following exercise. They found four trials with low sample sizes and a high-risk of bias. The authors concluded:

> There is insufficient evidence to determine whether whole-body cryotherapy (WBC) reduces self-reported muscle soreness, or improves subjective recovery, after exercise compared with passive rest or no WBC in physically active young adult males. There is no evidence on the use of this intervention in females or elite athletes. The lack of evidence on adverse events is important given that the exposure to extreme temperature presents a potential hazard. Further high-quality, well-reported research in this area is required and must provide detailed reporting of adverse events.

Another cryotherapy review[128], published a year earlier, concluded:

> There were no adverse events associated with WBC; however, studies did not seem to undertake active surveillance of predefined adverse events. Until further research is available, athletes should remain cognizant that less expensive modes of cryotherapy, such as local ice-pack application or cold-water immersion, offer comparable physiological and clinical effects to WBC.

These conclusions are tellingly unambiguous. An important distinction in the cryotherapy research, compared to reiki for example, is that studies are trying to elucidate the existence of real, physical mechanisms. In addition to perceptions of soreness, and subsequent muscle performance, we can also assess biochemical markers of inflammation and muscle damage, and determine if they're positively influenced by the intervention. It's considerably more difficult to cheat (or contaminate findings with placebo) when measures of physiological function are involved, and we should keep this in-mind when reviewing the data below. The research can also be differentiated into studies that are laboratory-based, and those that are conducted in the field (applied studies). While the former generally involve standardized exercise protocols, thus allowing a direct comparison between a cryotherapy intervention and the alternative, field studies comprise bouts of cryotherapy integrated into the athlete's regular training sessions which, by definition, aren't standardized. Given that a slight change from one training session to the next could heavily influence the dependent (outcome) variables, laboratory-controlled studies are considered more valid.

With respect to assessments on muscle soreness, one review[129] provided a breakdown of study quality and methodology (including the measures made and whether they were field- or laboratory-based). Of ten studies that had implemented muscle damage protocols to test the effectiveness of cryotherapy, only three had actually measured muscle damage *directly*, whereas the others relied exclusively on subjective reporting from participants. In these three studies, the relevant marker was creatine kinase (CK), an enzyme released from damaged muscle tissue into the blood, where it's measured with a simple blood sample; higher concentrations indicate more muscle damage. While one study found significantly lower concentrations of CK following five days of cryotherapy (administered twice per day), the two others found no differences in CK concentrations with three and six exposures, respectively. Being charitable, we might conclude that there's a dose-response to cryotherapy, and that benefits only manifest with twice-daily exposures repeated for five days or more. Interestingly, the lab trials do seem to consistently suggest a slight anti-inflammatory effect of whole-body cryotherapy, which wasn't observed with a no-intervention control. So, according to these data at least, cryotherapy might be a valid means of facilitating recovery when the alternative is passive rest. But let's contextualize these observations. A single cryotherapy session can cost in the range of £50–100 ($60–120). If the dose-response aforementioned really is our benchmark, ten exposures in five days seems an excessive expense, even for a professional athlete, especially considering the necessary time commitment and the numerous other treatments at their disposal. Moreover, a fully functioning chamber may cost a professional outfit hundreds-of-thousands.

Given the cost, the potential for adverse events that remains unexplored, and the evidence which is at best under-powered, lacking mechanistic underpinning, and dominated by studies of poor quality, I could not pragmatically stand by cryotherapy as a legitimate replacement for the many well-established recovery strategies that exhibit a stronger evidence-base.

9.11. Chiropractic

Chiropractic differs from many of the alternative therapies presently discussed because it has no roots in traditional Eastern medicine, and is a relatively new invention. The UK National Health Service provides the broadest definition of chiropractic: *a treatment where a practitioner called a chiropractor uses their hands to help relieve problems with the bones, muscles and joints*. The treatment was conceived in the late 1800s when D.D Palmer (an anti-vaccination proponent with a staunch belief in the magical properties of magnetic healing) allegedly restored hearing to a deaf man by adjusting his spine. Needless to say, the plausibility of such a miraculous feat is decidedly weak. Palmer, thus, concluded that nearly all physical ailments should be attributed to subluxations of the spine, and chiropractic was born. More contemporary chiropractic has distanced itself from subluxations of the spine as the sole determinant of illness, principally because it's an absolute claim that can be tested (via x-ray, for example). Chiropractic now self-identifies as a system of specific manipulations designed to free-up joints in the body that aren't moving properly; it's purported to influence neural pathways and general health.

Used extensively by individual athletes and sports teams, particularly in the United States, a 2002 survey[130] on chiropractic use among the teams of the National Football League (NFL) had 22 responses, and highlighted the sport's perceptions of this alternative treatment. Certified trainers generally considered chiropractors to have an important role in the NFL, principally for treating lower-back pain and other musculoskeletal injuries. A majority of trainers (77%) had referred at least one player to a chiropractor, and 31% of teams had one on staff. But such widespread use (at least within professional football) says nothing of the treatment's effectiveness.

While exploring databases for studies on chiropractic, it was apparent that the literature was saturated with case-studies, review articles, commentaries, and opinion pieces, but largely devoid of randomized, placebo-controlled trials (i.e., high-quality studies). This relative deficiency renders it problematic making any firm conclusions on the effectiveness of chiropractic beyond *we need better research if we are to recommend chiropractic*. Moreover, it's troublesome even interpreting some of these data given that such a majority are published in journals aligned with chiropractic; the *Journal of Manipulative and Physiological Therapeutics*, for example, is the official journal of the American Chiropractic Association. The *Journal of the Canadian Chiropractic Association* have a homepage on which they espouse the *scientific quality and vigor of the journal [sic]*; I suspect they mean scientific *rigor*, and the mistake doesn't inspire me on their rigor or vigor.

From a journal not directly aligned with a chiropractic organization came a 2010 study on chiropractic claims in the English-speaking world[131] with the aim of assessing the frequency with which affirmative claims were made about the effectiveness of chiropractic in treating asthma, headache/migraine, infant colic, colic, ear infection/earache/otitis media, neck pain, and whiplash. The authors reviewed 200 websites published by chiropractors, and the claims of nine chiropractic associations in Australia, Canada, New Zealand, the United States, and the United Kingdom. Their investigation revealed that 190 (95%) chiropractic websites made unsubstantiated claims regarding at least one of the aforementioned conditions. Moreover, 90% of websites and all nine associations made unsupported claims about headache/migraine, specifically. Such an overt disparity between claims and empiricism is a damning indictment and, as the study concludes, represents an ethical and public-health issue.

Although chiropractic-related adverse events are rare, there is documented evidence of gross negligence which warrants consideration, if only to serve as a warning on the dangers of *pseudo*science-based practice. Published case-reports and case-series highlight more than 200 patients who were suspected to have been seriously harmed during spinal manipulation, with the most common adverse event reported as vertebral artery dissections[132], whereby an artery located in the neck, and which supplies blood to the brain, sustains a flap-like tear to its inner-lining. While 200 injuries are more-than-sufficient, be mindful that this number represents just the published cases, and there likely exist countless others which remain unreported. Perhaps of greater concern is the practice of neonate chiropractic; the spinal manipulation of babies. There are several reports detailing injuries inflicted on babies and small children, including a 2013 case in which a baby's neck was broken by a chiropractor. The injury was first reported to the Chiropractic Board of Australia who dismissed the case without reporting it to the public, allowing the chiropractor to continue practicing on the condition he undertook education with an *expert in the field of pediatric chiropractic*. By way of defense, chiropractors have distanced themselves from these horrific scenarios, and are quick to label those responsible as *manual therapists* rather than trained chiropractors[133], while opponents argue they're one-and-the-same. Such pleading is reminiscent of the *no true Scotsman fallacy*, wherein one attempts to protect a generalization from counterexamples by changing the definition in an *ad hoc* fashion to exclude the counterexample.

> Geoff: No Scotsman would put ice in his whiskey!
> David: But my uncle is Scottish, and he puts ice in his whiskey…
> Geoff: Ah, but no 'true' Scotsman would put ice in his whiskey.

Despite the research generally ill-performed, and the negative press, not all chiropractic claims are wholly unfounded. Indeed, the use of chiropractic for the treatment of lower-back pain has some mildly interesting data. For some patients with lower-back pain, spinal manipulation by a trained practitioner seems to be effective, but not more-so than most other forms of treatment. Even in this narrow domain, the data are of limited validity due to methodological deficiencies. Indeed, a review[134] in *The Spine Journal* (the official journal of the North American Spine Society) highlighted: *No studies combine high-quality cost data with adequate sample sizes and controls for confounding factors*. From the American Chiropractic Association's own journal, a 2006 review[135] on the treatment of lower-extremity conditions (i.e., foot, ankle, knee, and hip) suggested that: *future chiropractic research should use higher-level research designs, such as randomized controlled trials*. The lack of RCTs, as aforementioned, represents a real deficiency in the data, and it appears that even for lower-back pain – which is the chiropractic mainstay – the evidence is decidedly weak. The principal concern is that double-blind studies (where both patient and practitioner are unaware of the treatment) are impossible, and single-blind studies are unlikely because patients generally know if their spine is being manipulated. Moreover, musculoskeletal pain is not an objective outcome. Finally, and mercifully, a critical review (a review of reviews) from 2011[136] collated data from no-less than 45 systematic reviews on numerous and varied conditions including: lower-back pain, asthma, pediatric conditions, and gastrointestinal problems. After assessing all outcomes, it was concluded: *collectively these data fail to demonstrate convincingly that spinal manipulation is an effective intervention for any condition.*

When it comes to clear and objective evidence, few practical and hands-on treatments can be considered to have such a negative record. And yet, chiropractic enjoys widespread mainstream success, with its own associations, clinics, degrees and doctorates, and many-a-practitioner on the payrolls of sports teams and private physiotherapy practices. It's sadly common for the commercial noise surrounding a practice to exceed the evidence in support of its use, but the mismatch between claims to efficacy and supporting evidence is as great here as I've ever seen. In fact, it's often the least-evidenced products that receive the greatest public lauding. This is something I've termed the *evidence-paradox*, and while its causes require more consideration, several potential attributions were postulated in *Chapter 1*. Many chiropractors are campaigning to be considered as primary caregivers alongside real physicians. This is a growing moral and ethical concern given the reluctance of some chiropractors to publicly acknowledge the deficiencies in the data, or take responsibility for the associated risks.

CAM is the result of therapies conceived without the grounding principles of evidence-based practice; at the very least, it's not underpinned by evidence that's deemed by modern medicine to be of sufficient standard, hence its alternative classification. Nevertheless, CAM thrives under such prohibition, and we see frivolous health advice administered in numerous facets of the industry. For example, consider *Goop*, a *modern lifestyle brand* founded in 2008 by actress Gwyneth Paltrow. The online magazine maintains an outwardly wholesome appearance, and the website catalogues articles on flattering styles of swimsuit, hair and makeup, food and fashion. Beneath the façade, however, numerous alternative practices are condoned, and while some are scientifically questionable, others may have demonstrably harmful effects. There are archives of articles on alternative medicine, supernatural phenomena, healing, and detox. Regarding detox therapies for the liver, for example, one article quoted an American naturopath:

> "I have found that the majority of diseases that manifest physically originate in the etheric energy body," … [the clinician] uses her own system

> of medicine. It draws on Ayurveda, homeopathy, yoga, anthroposophy, traditional Chinese medicine, and naturopathy to create personalized protocols to help people optimize their well-being.

Among the various concoctions available for purchase through the site is the *Implant-O-Rama System; At Home Coffee Enema*, priced at $135. This and similar products have been discussed in published medical reports documenting harmful repercussions of use (see *Detoxing* in *Chapter 6*). Despite being contrary to the advice of doctors and scientists – whose jobs depend on advising the public on best-practice in this respect – the product has sold very successfully to a global market. The MAYO Clinic (a prestigious, non-profit academic medical center based in the United States) advise that a colon-cleanse should only be considered when necessary to predicate a colonoscopy, and even then it should be performed under guidance from a physician, and by safer means. The *Goop* website famously published an article espousing the benefits of vaginal jade eggs, which was fast criticized by professional gynecologists. Finally, at the time of writing, the site contains a piece on *How to connect with people who have died*.

Many individuals (celebrities included) use their influence and reach to further their brand, help the needy, or promote a political message, but it's axiomatic that much harm can be wrought when such individuals offer advice on which they're not appropriately credentialed. It'd be particularly erroneous to prefer health advice from facets of the media over that from a physician or other health professional; the stakes are simply too high. We all have specialisms in which our opinions are valid; we should stick to our specialism. This critical advice is explored thoroughly in the final chapter.

10

CHECK YOUR EGO

op·ti·mism.

Hopefulness and confidence about the future or the success of something.

10.1. Free Yourself

In Freudian psychoanalytics, the *ego* is a constituent of the mind responsible for reality-testing, and which mediates between the primal and the cognitive psyches. In more conventional terms, the ego (noun) can be loosely described as a person's self-esteem or sense of importance, and it dictates how we perceive and interpret our world. Your ego may be large or small, in need of constant gratification or just the occasional reinforcement, and it operates silently in the background without instruction or permission. Your ego affects decisions and the way you learn. Most importantly, it's shaped by experience and, therefore, imprints a bias on all you see and do.

When the ego is challenged, even the humblest will strive valiantly to defend it. Consider ego and self-image as it pertains to social media. The online profile we spend such an inordinate amount of time crafting and shaping – whether you're the attractive, outgoing, and successful free-spirit, the confident and loving parent, or the brooding and compelling introvert – is the person we want others to perceive and the person we wished we were. Furthermore, if you're at all like me, you'll have found yourself on the wrong end of many-a-Twitter argument that seemingly has no beginning or end, and from which no reasonable conclusions ever result. Few believe that a forum underpinned by a series of back-and-forth 280-character pot-shots is an effective means of discourse; yet, we use Twitter to debate world politics, religion, science policy, abortion, vaccinations, and climate-change. The platform isn't distinct among social media, but many online discussions usually degenerate into an exercise in defending an ego-driven perspective and nothing more. Whenever mounting evidence exists to contradict an ideology we've been convinced was true, rather than perceiving this as an attack on our *ideas*, we perceive it as an *ad hominem* attack on our very being, and we take it personally. And if it transpires the opposing party were correct – self-evidently and objectively so – it can be disproportionately difficult to relinquish a position and risk damaging self-efficacy.

When it comes to acquiring new knowledge, challenging ego is of unconditional importance. In this respect, science is the foremost leveler because it cares not for preconceptions, opinions, or desires; truth – real, objective, measurable, demonstrable truth – is the sole commodity. In a 1964 lecture delivered at Cornell University, celebrated physicist Richard P. Feynman noted that knowledge and understanding come from testing your educated guesses against observations; crucially, Feynman said that if your guess disagrees with experiment, then it's wrong: *In that simple statement is the key to science. It doesn't make any difference how beautiful your guess is, it doesn't matter how smart you are who made the guess, or what his name is… If it disagrees with experiment, it's wrong. That's all there is to it.* This seminal quote is a reminder that no person, principle, or power is above scrutiny, and all should be subject to the same vehement examination. The ability to become emancipated from ego, therefore, is a critical prerequisite to intellectual honesty; only then can we maximize the likelihood of attaining true knowledge. Accordingly, this final chapter comprises various discussions and meditations on how to enact this, and the broader implications. Moreover, although this has been a book about critical-thinking in the health and fitness industry, I presently aim to collate the various themes pervading it, in order to consider them in a broader context. I'll also provide some closing thoughts and ideas for future direction.

10.2. Yearn to Learn

Slouched on a park bench one sunny Saturday afternoon, watching the world pass by, I spy two pigeons nod their way along the path in front of me. The bird on the left exhibits classic gray plumage with bright, metallic green feathers on its neck. As it trots along, the feathers become iridescent, shimmering and pulsating in reflecting the sun's rays. Its counterpart looks identical in size, shape, and color, except the feathers on its neck are reflecting an equally brilliant metallic purple. I wrongly assume the first pigeon to be female and the second to be its male mate. However, as the second bird moves along, the purple on its neck transmogrifies to match the jade-like green of its comrade; on both birds, therefore, the colors are constantly changing with their horizontal movement. My overwhelming response was one of intrigue, and I actioned immediately to learn more of the mechanism – the physics – at play to evoke this fascinating color change. Perhaps more pertinently, I wanted to learn what purpose it served, i.e., the evolutionary pressure which caused these pigeons to exhibit such an interesting characteristic.

Unsurprisingly, this is a widely studied phenomenon. In fact, an article from 1902 in the journal *Biological Bulletin*[137] keenly described the mechanism in a paper entitled *The Metallic Colors of Feathers from the Neck of the Domestic Pigeon* (go figure!). After studying their appearance with the naked-eye and then under a microscope, the author (R. M. Strong) concludes: *The greenish effects are produced when light strikes the broad surfaces… and is reflected with a small angle of reflection. The reds appear only when light falls with a large angle of incidence…* With respect to the *why* question, the iridescent feathers may function to attract a female mate and, as with most other birds, the male pigeon exhibits greater iridescence than the female.

I've always been fascinated with seemingly banal subjects, because it's these that often provoke the most intriguing questions. I recall a phase during which my young nephew (at the time, he was only four years old) greeted every phenomenon be it natural or supernatural with a *why* question; why are the trees green? Why is there wind? Why do magnets pull? These are such simple questions with not-so-simple answers; have you ever attempted an explanation of magnetism to a four-year-old? Like all children, we are born into ignorance, but with an innate yearning to understand and assimilate the world around us. We have an engrained sense of wonder and inquisitiveness. As we grow, this innocent naiveté is lost and often replaced with a general overconfidence in one's own perceived knowledge, a syndrome coined by and named after Dunning-Kruger. But such a sense of wonder must be nurtured and maintained through adolescence and into adulthood.

Sadly, this thirst-for-knowledge is sometimes absent in young adults. I've taught a number of subjects at various academic institutions, but I'm principally interested in exercise physiology. I've run successful applied physiology classes with around 120 students (yearly) in which I've given, as reading tasks, 15–20 landmark journal articles spanning the last several decades of physiology research. If you're an exercise physiologist, you'd likely be able to guess at least ten of those on the list, because they're papers that informed and changed our perspectives on key aspects of the discipline. Two examples are the seminal work by Hill and Lupton in characterizing maximal oxygen uptake in man, and the early research on gas exchange threshold by Karl Wasserman and Brian Whipp. In any case, they're principles of which accomplished undergraduates should have a firm grasp. These are among numerous systems in human physiology that, to my mind, cannot fail to stimulate a sense of awe and intrigue in the hearts of those fortunate enough to study them, and they're a joy to teach. I use such fascinating mechanisms as vessels with which to inspire and instill within my students a raw fascination with the basic operations of life. Moreover, as an educator, it's an integral part of my remit to do so. We must do all we can to inspire the next generation with the appetite for wonder that's essential for life in the modern age. From his book, appropriately titled *An Appetite for Wonder*, Richard Dawkins reflects:

> The purpose of a lecture should not be to impart information. There are books, libraries, nowadays the internet, for that. A lecture should inspire and provoke thought.

To convey how such an appetite for wonder might be nurtured in the present context, I'll borrow a Chinese idiom which was originally coined to facilitate learning in the martial arts: *don't consider yourself strong, but consider yourself weak!* The premise behind this cryptic phrase is as follows. Martial artists train their entire lives in various forms of self-defense, or fighting styles, striving to harden their bodies and minds. As such, they sometimes tread the path to overconfidence and arrogance. Such bravado breeds complacency, and this state-of-mind may lead someone into situations they'd have otherwise avoided, e.g., walking down a dark alley at night, or through a rough neighborhood. If one considers themselves as weak, however, they'll take all of the necessary precautions to circumvent risky situations, and mitigate the likelihood of a confrontation in the first instance. The credo is that one should strive to become strong but, to avoid the pitfalls wrought of arrogance, one should behave as though they're weak. If we adapt such a philosophy for the present context, the moral should be evident. We should strive, as best we can, to educate ourselves on the nature of the modern world, encompassing, for example, science, history, literature, and the nuances of critical-thinking. In doing so, we strengthen our minds and inoculate ourselves against the exploitation of our ignorance (i.e., we get strong). Nevertheless, to avoid the overconfidence that can lead to an incorrect assumption or inhibit the acquiring of new knowledge, we must also remain humble, inquisitive, and not assume we already have the answers to life's important questions. Assume that you're ignorant, naïve, and remain always amenable to learning new things (i.e., consider yourself weak). Essentially, in order to safely navigate the health and fitness industry, and the broader landscape, with all its snags and snares, we should think like a toddler, foster that *yearning for learning*, be inquisitive, and keep asking *why*.

10.3. It's OK to Defer

In questioning the truth of a matter (e.g., does this product work, is this a reasonable mechanism-of-action, will this program help me lose weight, will this drug accelerate my recovery?), there are two broad approaches we must take. First, as alluded to earlier, we must try as best we can to educate ourselves on the subject specifics. If you're buying a supplement, then conduct some research into the various brands and the evidence-for-efficacy; if you're commencing a new training program, then speak to some trainers, look at any published literature, and ask your friends who might have trialed it directly; if you're beginning a new dietary regimen to lose weight, check to see if it provides you with the nutrients to meet your individual demands, and whether it's long-term sustainable. Claims in the health and fitness industry are generally associated with changes in the body one might expect to see following a new intervention. The claims are often specific and testable, and reference supposed mechanisms of the body. So, using validated resources, do some study on those reputed mechanisms, and establish if the claims are reasonable given the state of knowledge. Acquiring a rudimentary understanding of *how things work* will help provide some protection against products aimed at exploiting ignorance, and this will make you less of a target. In terms of how best to do this, it'll depend largely on the scenario. Each of us exhibits an individual capacity for learning, and responds differently to various stimuli. It will take time (and perhaps trial-and-error) to establish the means by which you learn most effectively. For me, reading is essential, as much and as often as possible. However, reading passively can often be uninspiring. Instead, I deface pages of text with a red marker (pencil if it's a book…), scribble notes, draw pictures, connect ideas with circles and flowcharts, and summarize key paragraphs in concise sentences. I'll fold page-corners of particularly pertinent sections. Engaging with the material in this manner (i.e., more actively) enables me to process it more effectively. Independent of reading, I listen to audiobooks and lectures, when I drive and when I run. Far from a burden, a long car or train journey is a gift – the gift of time – affording you the solitude to become immersed in a complex mechanism, a biography, or a story. While the notion that each of us responds best to a certain learning style (e.g., audio, visual, kinesthetic, etc.) has been widely debunked, I find that combining several methods of information processing is the most effective. Accordingly, I'll read, write, make notes, watch videos, listen to lectures, discourse with colleagues, and design short presentations on those topics I've studied. But what works for me might be ineffective and uninspiring for you. So long as you access information from a verified source, that's representative of the current understanding, then explore the various means at your disposal. The key is to make a sincere effort to upskill, taking ownership over the decisions you make by ensuring they're educated and informed.

By contrast, almost paradoxically, we simultaneously must develop a keen sense of when our understanding is left wanting, and otherwise insufficient to allow us to make an informed decision. Sometimes, there just aren't enough data available, or perhaps you don't have the time to do a *deep dive* on the literature, or maybe the subject is just outside of your capacity to understand. In the latter regard, I'll likely never understand quantum physics at any real depth. Irrespective of the books I read and the conversations I have, I fear that my brain isn't wired in such a way that permits me to visualize a fourth dimension. As a result, there's a tangibly modest limit on the extent to which I can understand space-time and general relativity. That's OK with me and I've (sort of) made my peace with the fact. It's crucial that we learn to acknowledge our shortfalls. Doing so can be exceedingly difficult, especially in the modern post-truth context in which everyone is an expert. Nevertheless, when you're insufficiently knowledgeable on a subject to be confident that you've landed on the correct answers, it's necessary to defer to an expert who might have: a real, authentic, credentialed expert. At the very least, be willing to concede to someone whose knowledge surpasses your own. By all accounts, there's nearly always available an authority who knows more than you about a given subject. Moreover, one should only arrive at *definitive conclusions* when the majority of experts agree by consensus.

Much has also been said about *elitism* in science and policy, and it's important to contextualize this notion in the present discussion. Elitism – the belief that a society or system should be led by an elite – isn't necessarily a bad thing. On the one hand, it's associated with a sort of condescension toward people who don't share a similar status. Elitism in the context of personal or class superiority, characterized by a certain lifestyle, education, wealth, and privilege, elevating one above others, is the negative aspect of the philosophy. On the other hand, if you snapped one of the fragile bones in your foot playing football or hockey, and needed surgery to fix the bone in-place using metal screws, I'm sure you'd be acutely mindful of the skills and experience and status of the surgeon afforded the task. In the case of open-heart surgery, the notion of employing an elite surgeon isn't an offensive one. You'd hope that the person piloting your colossal crew vessel wasn't a teenager on work experience during her summer vacation. One would expect the commander of a space station, responsible for billions-of-dollars of technical equipment, to be appropriately credentialed, having undergone something close to several decades of arduous training, most likely at the head of their field. When the stakes are sufficiently high, none of us seem to rebuke the notion of elitism; indeed, we'd expect it. Imagine a world in which only medics and scientists were allowed to regulate vaccination policy; only career environmental scientists (whose jobs depend on a correct interpretation of climate-change) were able to dictate the research and funding into renewable energy; only elite nutritionists were given the responsibility of regulating sports supplements; only elite physiologists with their superior understanding of human metabolic function were tasked with designing weight-loss programs alongside elite psychologists at the forefront of elucidating the complex relationship that we have with food. Such a world would have fewer communicable diseases, be closer to renewable sources of energy, a few dietary supplements that worked, a healthy population that met their physical activity needs and who followed a dietary regimen that was long-term sustainable, and an obesity epidemic rapidly attenuating in response to people's improved understanding of their emotional connection to food. We'd have fewer extravagant claims, more rigorous evidence-for-efficacy, and a system that was unimpeded by narcissistic self-obsession. Give this due consideration the next time you believe all opinions to be equally valid. The simplest solution to a problem isn't always simple, and it sometimes requires an expert on-hand to formulate the most appropriate response.

Striving for education might appear diametrically opposed to deferring to expert knowledge, but it's rarely one or other (it's a false dichotomy). To stand the best chance of navigating this complex world, we must learn to holistically integrate the two perspectives. In other words, aim to be knowledgeable on topics which influence your life, or on which you would like to have a particular informed opinion, but be willing to defer to *expert opinion* on that and other topics, assuming all of the caveats aforementioned. This is the most direct path to elucidating truth.

10.4. Knowledge of Ignorance

Let's further explore this notion by returning to Feynman; he once said of critical-thinking that *The first rule is that you must not fool yourself, and you are the easiest person to fool!* Feynman knew all-too-well the dangers of self-deception in science. Indeed, one of the foremost barriers we face in furthering our knowledge and understanding of the world is acquiring the capacity to admit our own ignorance of subjects, even those about which we're supposedly educated. This is no minor task, for even the humblest among us might protest when their understanding or intellectual credentials are brought into question. Nevertheless, admitting a degree of intellectual fragility is the first step to strengthening that fragility and avoiding erroneous claims.

While a mention of Socrates – the founding-father of the scientific process – is somewhat delayed, one of his philosophical teachings is of particular relevance here. Among many other initiatives, Socrates is known for coining the Socratic Method. At its core, the Socratic Method is a form of cooperative discussion between individuals, or among a group, predicated on asking and answering questions to stimulate critical-thinking and to evoke new ideas and underlying presuppositions. Essentially, it can be considered a system of reasoning and rationalizing a subject in order to better understand it. In such a discourse, hypotheses are sequentially proposed and discussed, and discarded if they fail consistent tests of logic. The outcome is a general, commonly-held truth on which all agree. The civilized world has a pretext of being underpinned by a Socratic Method; in it, all actions are discussed, reasoned, rationalized, and agreed-upon by all-parties. Such a world-wide institution is, perhaps, an unrealistic utopian fantasy. The Socratic Method was the basis of scientific investigation in Ancient Greece, later combined with the objective testing of observations or experiment as soon as humans developed the means. The process of Socratic enquiry is underpinned by another of his teachings; that humans must be knowledgeable of their own ignorance.

The Socratic Ignorance (aka. Socratic Wisdom) .is, arguably, one of the Greek philosopher's most famous legacies, and it's a perspective associated with humility that all should strive to emulate. In Plato's *Apology* – one of the most lauded of philosophical works – an account is given of Socrates' appearance in court on charges of impiety. Indeed, he was considered to be corrupting the youth, a crime for which he'd eventually be condemned to death. The speech he offered in his defense recounts how a villager – Socrates' close friend – was informed by the Delphic oracle that Socrates was the wisest man alive, an accolade that Socrates himself refuted. In an attempt to prove wrong the oracle, Socrates ventured into the city to discover people who displayed exemplary knowledge and skills that far-exceeded his own (e.g., cobblers, writers, blacksmiths). But on discovering these talents, it emerged that they also thought themselves supremely educated in all-manner of things, which was evidently untrue. Socrates, it seemed, was the only individual knowledgeable of the true reach of his own ignorance, and he eventually conceded: *I am the wisest man alive, for I know one thing, and that is that I know nothing.* Further to this account, the writings of Socrates are littered with scenarios in which he openly acknowledges his ignorance of the topic under exploration.

Such knowledge of ignorance is a prerequisite to learning. Consider the means by which some of the major discoveries of the past several centuries were made. Galileo, the Italian astronomer, was a proponent of the heliocentric model of the solar system, which is the proven proposition that the earth and other planets orbit the sun. While the Roman Inquisition admonished the idea as a *foolish and absurd in philosophy, and formally heretical since it explicitly contradicts in many places the sense of Holy Scripture*, Galileo refused the traditional biblical hypothesis and, using an early form of telescope, made direct observations of the celestial objects to support his position. For this, he was placed under life-long house arrest for heresy. Until the late 1800s, the principal cause of physical ailment was an imbalance of bodily *humors* to include blood, bile, and other fluids. This antiquated and now disregarded view resulted in the practice of blood-letting in which a patient's blood was drawn to restore balance. George Washington famously requested to be *bled heavily* after developing a throat infection in 1799. Nearly four liters of blood were supposedly withdrawn in a 10-hour period prior to his death. Although still practiced in some particularly primitive forms of medicine, advances in modern science alongside the germ theory of disease – by, among others, Louis Pasteur in the 1850s – led to blood-letting being cast aside in favor of more effective, less dangerous treatments. In the sporting domain, lactic acid was demonized for many decades by physiologists intent on over-simplifying the cause of muscle fatigue during high-intensity exercise. Our lack of understanding, coupled with the yearning for a single, overarching explanation for our inability to tolerate prolonged forceful muscle contractions, led to the incorrect assumption that this organic acid inhibited muscle contraction. In the 1980s, Professor George Brooks coined the term *lactate shuttle hypothesis* which described the movement of lactate within and between cells, thereby reinforcing the notion of lactate as a fuel source during exercise. Collectively, these choice examples depended on career scientists not accepting the models which had been assumed to explain complex phenomena of the time. Only when it's evidence-based and reinforced by multiple lines of enquiry converging on a factual basis, can we be confident in the authenticity of a proposed paradigm. For all other questions, honest enquiry must follow an authentic attempt to accept that one doesn't already have the answers.

10.5. Be Ready (and Willing) to Change Your Mind

Being the owner of sound knowledge and solid reasoning skills lowers the likelihood that you'll defend an incorrect position, but checking your ego is critical to this process given that you'll be more willing to change your mind on the strength of evidence, and accept the world from another point of view. Ideological positions – a set of established/staunch opinions or beliefs of an individual or group – aren't usually established on evidence. In fact, they're usually retained because they align with political, cultural, religious, or other tribal perspectives. If such a view is able to be reconsidered on the weight of existing evidence, then the view ceases to be ideological. But reason is an evolved trait which provided early humans a survival advantage by helping them to win arguments in hyper-social groups, and it's less well-adjusted for establishing truth, *per se*. Moreover, reason is painfully limited in myriad ways. For example, a fact of human nature is that we, as a species, like to cling to traditions. If you believe that blocked energy channels in the body cause pain and disease, then you do so irrespective of the fact that there's no evidence on which all can mutually agree. If you have an absolutely unerring faith in this self-perceived *truth,* and you're not open to potentially changing your mind or taking another perspective regardless of the contrary arguments with which you might be presented, then you are defending an ideology, plain and simple. When such is the case, you're more likely to accept and assimilate evidence that confirms your predetermined beliefs, and simultaneously rubbish and reject that which contradicts them. This is just as true for your beliefs pertinent to politics and religion as it is for your beliefs in HIIT, low-carb diets, reiki, yoga, or energy medicine. It's also true that facts and figures, however objectively true, often aren't enough to break the shackles of an established system of thought.

A seminal paper to study our innate ability to log evidence that confirms our beliefs, and dismiss that which contradicts them, was first published at Stanford University in the late 1970s[138]. A cohort of undergraduate students were divided into two groups of equal size. The first harbored a belief that capital punishment (the death penalty) was a positive system for deterring crime; the second had the opposing view that the government-sanctioned death penalty had no effect on crime and should be abolished. Each student in the cohort was given two studies to which they had to respond; while one of the studies provided robust data that supported capital punishment as a deterrent, the other study provided equally robust data that contradicted the first. Both studies were completely fabricated, designed for the purpose of providing compelling, and comparable, statistics on the opposing view. As one might expect, when responding to the two studies, students in favor of capital punishment found the study which confirmed their beliefs to be robust and credible, whereas the opposing study was perceived to be methodologically flawed and relatively easy to dismiss. Those students against capital punishment arrived at precisely the opposite conclusion. When asked again about their views on capital punishment, both groups had diverged even further, representing their pre-existing beliefs more staunchly. This study has been replicated and expanded upon for many decades, and the findings show consistently that confirmation bias (a pre-existing belief) is the spanner in the works of any system founded on logic and reason.

Given the abovementioned research, let's briefly revisit a subject discussed in *Chapter 6*, the misinterpretation of the literature pertaining to alcohol consumption. This led to the incorrect advice that drinking a glass of wine-a-day conferred greater health outcomes than complete abstinence. Instead, when accounting for the fact that the abstinence groups that were studied included individuals who abstained due to ill-health, the benefits of moderate consumption disappeared. There's now first-hand evidence from a very robust study published in the journal *Nature*[139] that drinking alcohol can cause irreversible damage to DNA, including to stem cells. Prof. Linda Bauld, a cancer-prevention specialist from Cancer Research UK which partly funded the research, commented: *We know that alcohol contributes to over 12,000 cancer cases in the UK each year, so it's a good idea to think about cutting down on the amount you drink*. The study hasn't received widespread coverage, nor was it printed on the front-page of mainstream media outlets. Could it be that the research, however compelling, has been largely overlooked because it conflicts with an established narrative? Certainly, the flawed conclusions on the benefits of moderate consumption received far greater exposure than the subsequent reports correcting them. Many people live their lives on the precept of *everything in moderation*, and become unsettled when confronted with evidence forcing a rethink of established norms. The reality is that some things can, and should, be excluded from our lifestyles because they confer us no benefit. It's likely that both alcohol and refined sugars fit this category rather well. If you choose to drink, do so because you enjoy it, and not the misapprehension that it's good for you. We must foster the intellectual capacity to let go of old and misinformed ideas, however entrenched, when they're contradicted by evidence.

Ironically, it would seem, it can be ineffective teaching people to form opinions on the basis of evidence because we harbor inbuilt mechanisms of bias which see us justifying the dismissal of evidence that contradicts our beliefs. While this can be partly addressed by teaching people how to recognize *good* evidence, in addition to raising the standards of what's considered *robust* data, it's often insufficient. Directly challenging a staunch belief will likely provoke a defensive response because nobody likes to be challenged. But it can be more effective to encourage an individual to reassess their own beliefs, on their own terms. Peter Boghossian at www.skeptic.com suggests:

> Instead of telling people to form beliefs on the basis of evidence, encourage them to seek out something, *anything*, that could potentially *undermine* their confidence in a particular belief… this makes thinking 'critical'.

Boghossian discusses the process of *defeasibility*, which is when somebody deliberately searches for evidence that might undermine their pre-existing beliefs, and it's a valid means of determining the extent to which such a belief is revisable. For example, in the preceding chapter, I provided an overview of the research pertaining to chiropractic. The direct evidence for the effectiveness of chiropractic is poor, and scarcely extends beyond a series of *it worked for me* anecdotes. Assume that Sarah is a long-time proponent of chiropractic, and believes that regular visits can relieve shoulder pain, headaches, and stress, and help her train more effectively. An attempt to persuade her otherwise – by presenting facts and figures according to the research – would be squandered time for you both. The more entrenched her belief, the more futile your efforts. Ask Sarah to consider what it would take to shift her, even slightly, from this seemingly immovable position. What if she spoke to a friend who'd had a bad experience with chiropractic, or if the bad experience were her own, or if her own chiropractor conceded that the evidence-for-efficacy was poor? Perhaps Sarah would then begin to question her belief in chiropractic as an alternative-therapy cure-all, or maybe she'd resoundingly dismiss the negative press. The outcome is that you become enlightened as to the extent to which her ideas are fixed or finite. Defeasibility and similar exercises are useful because they cleave a space in someone's mind for *doubt*; that crucial word, integral to open and honest enquiry, that detaches us from unwavering faith.

I subjected my own steadfast view on sports tape to a test of defeasibility. This widespread practice sees athletes in sporting competitions around the world with strips of fluorescent tape strapped to various body parts including the legs, back, and shoulders. The practice has wormed its way into pregnancy *pseudo*science, and expectant mothers can often be seen with tape and strapping along the contours of their abdomen. Companies claim that the tape can enhance performance, improve blood flow, reduce injury, and reduce muscle soreness, but there's no good evidence to support the assertions. What little data exist suggest that specialist sports tape is no more effective than other brands in the management and/or prevention of sports injuries, and that sports tape is placebo tape. It bestows athletes with the impression that they've received a functional intervention, thereby boosting confidence and altering their perceptions. Under what circumstances would I alter my stance, and what would need to happen before I'd perform a complete 180° about-turn on the subject? Principally, my position on this isn't faith-based or unfalsifiable. Something unfalsifiable is asserted and assumed to be true, even though it cannot be quantified or contradicted by an observation or any physical test. Unfalsifiable claims are not based on logic, reason, or evidence. My stance *is* falsifiable. It can be challenged, tested, and proven erroneous under appropriate conditions. In answering the aforementioned questions, all that I'd require to change my position is some strong (high-quality) evidence, from independent labs, not funded by the manufacturer, and that were sufficiently powered and scientifically rigorous to overturn the existing glut of poor, low-quality data. In fact, as with all topics on which I've established a perspective, I'm constantly challenging my stance based on the most contemporary evidence, as are all good scientists. If opinions are formed on the best available evidence then, by definition, those perspectives will be constantly evolving with the times. Critically, I present not *my* opinions; I present object facts of scientific consensus. I am not personally or emotionally attached to them, and if proven wrong, and all agree on how *proven wrong* should manifest, those perspectives will be unceremoniously discarded. All subjects should be met with the same objective indifference. Can you imagine a proponent of reiki, cupping, or chiropractic expressing an equivalent dispassion if their views were to be challenged and disproven?

Someone dismissing evidence that is considered a threat (solely on the basis that it contradicts their position) is the cognitive equivalent of burying their head in the sand. Yes, it's an innate human characteristic, but making no effort to correct for such bias shows a real degree of intellectual paucity. The mere fact that humans can be critical of someone's viewpoint – seeing the apparent flaws in their reasoning – while concurrently being blind to their own intellectual insufficiencies, is evidence of the extent to which our cognitive capacities can be limited. These are skills which, like any other, must be trained and honed, at great effort and expense. It won't happen overnight, but it will happen. And while changing your mind and taking the world from another perspective might be a perceived sign of weakness, doing so goes against our natural inclinations and, therefore, actually requires great strength. Whether you're a chiropractor, a yogi, a paleo proponent, a vegan, a flat-earther, a scientist, a republican, a democrat, or a Brexiteer, good critical-thinking begins by accepting the notion that any given belief about the world *might be wrong*. More importantly, there must be a scenario one can envisage in which they'd revise that belief if appropriately contested and disproven, regardless of the damage sustained to ego or self-efficacy.

10.6. A Note on Arguing with Others

Although the themes pervading this book prioritize sports products marketed at the consumer, it's likely that you'll disagree frequently with the perspectives of others on these issues, and may at some point find yourself in dispute. It's, therefore, worth noting several prerequisites to arguing with others. In some arguments, you'll be the provocateur, and in others, you'll be on the receiving-end of criticism. Sometimes, a confrontation reflects a reasonable and civilized cooperative testing of hypotheses, with the joint aims of establishing a shared *truth* (as with the Socratic Method). Other times, it's a series of vitriolic *ad hominem* attacks to discredit your adversary and keep self-efficacy intact. In any case, it's worth noting just a few good habits that'll help you keep your ego in check, and maximize the chances of drawing reasonable conclusions from a discourse.

Be charitable and generous with respect to your opponent's position. This doesn't mean naively trust their motives which will sometimes prove malevolent. Rather, to stand the best chance of understanding (and perhaps refuting) their arguments, you must view the argument from their perspective, even if it means temporarily crediting them with benevolent intentions. Make a sincere and concerted attempt to empathize (or at least sympathize) with their position. You should also strive to understand their argument, with all its nuances, because you cannot oppose an argument you simply don't understand. Study the evidence in favor of their position, and consider generously the logic they use to justify it and formulate their conclusions. Such an altruistic motive is integral in progressing the argument toward a meaningful conclusion. Moreover, you need to be confident that any conclusion is valid (on merit) and not simply the result of your superior debating skills. Arriving at an objectively valid truth should be the priority, valued above all else, even if achieved by conceding your argument and seeing another in its place. Accepting this as a prerequisite of the process will spare you much anger and frustration. It requires that you practice the very challenging art of *checking your ego*.

In his bestselling book *The Chimp Paradox*, Prof. Steve Peters proposes a model for managing the spontaneous and emotional behaviors that can lead to negative outcomes, but I'll focus on the propensity for such behaviors to sour logical discourse. Peters gives his readers the opportunity to learn a new skill, that is, the ability to *choose* how to respond to a given situation. His model, a simplified representation of various facets of behavioral psychology, proposes the *human* (the version of yourself with whom you identify and would most like to be associated), and the *chimp* (a potentially destructive primate that's not under your direct control). The two components of the psyche differ markedly. Chimps are primal, strong, and notoriously unpredictable, whereas humans are considered the *rational animal*, capable of calm and logical discourse. When a chimp argues, it does so emotionally, aggressively, unconcerned with establishing mutual ground. And while the chimp acts of its own accord, you are (of course) responsible for managing it and ensuring it isn't overly destructive. This can be troublesome because chimps are far stronger than humans. Nevertheless, Peters proposes strategies to successfully nurture and manage your aggressive primate cousin. Pertinently, other people also have chimps, acting and reacting on their behalf. When two people engage in a discussion on *any* given topic, it's critical they're both able to *cage their chimps* and debate reasonably using their human brains. If the chimp of one individual makes an appearance – with an aggressive stance, defensive position, or a hurtful comment – it's likely going to provoke the other. Being improbable that either chimp will yield, the argument escalates until both chimps are beating their chests and flinging feces in each other's direction. Once a situation is grasped with emotion, it becomes almost impossible to diffuse. Nurturing and managing your chimp (and mitigating the likelihood of it manifesting in an argument), as well as understanding and placating your colleague's chimp, is crucial for seeing constructive outcomes to any argument. Peters suggests:

> Chimps that are insecure may read lots of things into harmless situations. They can also read intrigue and malice in comments or statements that others make and then allow their imaginations to run wild... The Chimp does not necessarily work with facts but it works with what it believes is the truth or with a perception of the truth or, even worse, with a projection of what might be the truth. It is quick to form an impression on little, if any, evidence and usually won't give way. Of course, some impressions that the Chimp gives us are accurate and helpful, but they can just as easily be wrong. Searching for some accuracy and truth would help us to reach a sensible conclusion.

The chimp model is pertinent here because avoiding erroneous conclusions depends very much on our ability to think clearly, objectively, free from contaminants like bias, speculation, and emotion, all of which are fuel for the angry chimp. *The Chimp Paradox* describes how to find common-ground between our logical, rational cognitive brains, and our primal, spontaneous ego-driven chimps to stand the best chance of a happy and agreeable outcome.

10.7. Future Directions

When compared to other disciplines, the sport and exercise sciences (to include physiology, nutrition, sports psychology, strength and conditioning, biomechanics, physiotherapy) are young endeavors. Because of this relative infancy, some of our guiding principles are less well-established, in both academic and applied circles. When coupled with the vast commerciality of health and sport, one arrives at an industry built on the exploitation of a relative paucity of public understanding. What needs to change, therefore, in order to improve the state of the commercial health and fitness industry?

We must read more. Reading books and journal articles (at least, reading the right ones…) is a principal means of assimilating new information, understanding how to think critically, becoming educated on complex issues, improving vocabulary, learning how to think, speak, and write articulately, and learning how to distinguish between information and misinformation. Sadly, dedicating spare time to reading for pleasure is not a characteristic of contemporary culture. Surveys of over one million U.S. teens collected since 1976[140] suggest a substantial rise in the recreational use of digital media (including social media, gaming, texting, and using the internet), but a large decrease in the use of contemporary media (such as books, magazines, newspapers, TV, and movies). Moreover, while data from the 1980s show that 60% of 12th-graders read for pleasure, estimates for 2016 were as low as 16%. Although we can't conclude cause and effect, it's unlikely that the rapid rise in the use of the internet and social media is independent of this phenomenon; as the authors wrote in an article published in *The Conversation* (2018):

> Of course, teens are still reading. But they're reading short texts and Instagram captions, not longform articles that explore deep themes and require critical thinking and reflection. Perhaps as a result, SAT reading scores in 2016 were the lowest they have ever been since record keeping began in 1972. It doesn't bode well for their transition to college, either. Imagine going from reading two-sentence captions to trying to read even five pages of an 800-page college textbook at one sitting. Reading and comprehending longer books and chapters takes practice, and teens aren't getting that practice.

Twitter feeds contain *threads* of individual *tweets*, each a maximum of 280 characters (increased in 2018 from 140). Many longform writings that exceed this modest character limit are likely to be sacrificed in a climate where we've become conditioned to instant gratification. Students often struggle to complete complex reading tasks that underpin their assignments, and such longform deconditioning is a likely culprit. As the aforementioned authors suggest, acquiring the concentration necessary to assimilate longform text requires time and practice, and students aren't getting that practice. The preference for easily digestible, bitesize chunks of information has led to several physiology and medicine journals adopting short-form articles (typically around 800 words) which are more amendable to the busy lives of medics and academics and, thus, more likely to be read. Novel initiatives like this might curb the pervasive practice of learning study outcomes only via the published abstract, and it might be a means of inspiring students to healthier reading habits.

Broader education in critical-thinking. We've established that critical-thinking is integral for all vocations, trades, and academic pursuits; yet, it's absent from the curricula of many high schools, and not always very accessible to adults who fail to see its relevance to modern living. It's my hope that books like this help to readdress the focus, but more needs to be done. Critical-thinking should be taught more broadly in courses at high school and college/university. Most students of the sciences take classes in Research Methods (a study of the qualitative, quantitative, or mixed tools used in research) or an equivalent, but critical-thinking must be studied outside of this narrow domain. If we widen the scope to see its practical relevance in medicine, business, commerce, finance, creative arts, and politics, then we stand a better chance of developing graduates with these skills embedded more holistically. They will, therefore, be better equipped to navigate the world regardless of their field of study or chosen career-path.

More emphasis on good science, and less emphasis on doing research for career-development, monetary gain, or notoriety. Many of my colleagues in kinesiology/sport and exercise science unwillingly end up on the academic treadmill. This happens when focus drifts from academic curiosity and the raw pursuit of knowledge, and toward the never-ending game of scholarly top trumps that serves the primary purpose of fueling egos and pay checks. Idioms like *publish or perish* are damaging; they provoke within academia an urgency to publish at all costs, regardless of the relevance, standard, or rigor underpinning the work. When profile and the next promotion become master, it's often the quality of the science and/or the effective communication of the message that's first sacrificed. This scientific integrity is particularly fragile when there's additional pressure from external investors, or when studies are funded by corporations who have a vested interest in the outcomes. In 2018, Alexander Clark and Bailey Sousa, for the Times Higher Education (UK), wrote:

> Academic work is difficult. It is often isolating and highly competitive. It is distinctively demanding of our capacity to make choices – and of our intellect, emotions, skills, energy and creativity. And it is perilously boundless: we can never teach well enough, publish enough or get enough funding. Research and teaching pressures increase and diversify as job prospects, pensions and pay deteriorate. Long working hours and lack of life-work balance seem like the new norms. These not only harm mental well-being but also increase workplace disengagement, bullying and even questionable research practices.

In 2012, a Harvard study[141] surveyed over 2,000 psychologists regarding their involvement in questionable research practices, and provided a clever incentive to increase the chances of honest responses. The study found that 78% of respondents admitted minor offenses like failing to report all dependent variables, while

72% admitted collecting more data after assessing whether the initial results were significant. The context in which this latter infraction was committed deserves consideration. If, for example, the researchers performed a power analysis and established in advance that they needed 20 participants, but ran the statistics after they'd collected data from only 15, they might be forgiven for trying to avoid additional time and expense (although, certainly, power analyses are performed for a reason). If, however, there was no power analysis, data were collected from an arbitrary number of subjects, and researchers performed statistical tests periodically until observing significance, this is known as *p-hacking* and is decidedly poor practice. More concerningly, 39% admitted rounding down their *p* values (which gives the impression of stronger significance), and 9% admitted falsifying data. This last point is no minor transgression. Indeed, an undergraduate student caught falsifying data would be taken to an academic misconduct panel, and likely expelled. Moreover, it's worth remembering that while the study offered incentives for honest reporting, the survey results likely underestimated the prevalence of questionable research practices. Interestingly, the study found that respondents who admitted to a questionable practice tended to think that their actions were defendable, and this was independent of the area of research or discipline. Less severe misconduct (e.g., not reporting all dependent variables) is easier to justify, whereas the more severe violations (e.g., falsifying data) were less so, but it's a telling insight into the ways that humans rationalize their poor decisions. It should be noted (before I receive emails) that while this study surveyed psychologists, these findings are likely much the same, or worse, in other scientific disciplines. The prevalence of poor research integrity was attributed to publication pressures, motivated reasoning, and professional ambitions, which collectively manifest in all scientific disciplines.

So, according to this study at least, there has to be greater emphasis on good science practice for the sake of good practice. Scientific rigor doesn't flow from industry investment, publication pressures, career aspirations, or any other form of extrinsic motivation. The money now associated with high-impact publications serves to reinforce these factors in a loop of positive feedback. But while peer-pressure to flourish in academia – an inherently competitive domain – can sometimes negatively impact on rigor, peer-pressure can also serve our cause. By holding our colleagues to account for their research, taking more time to review their work, even scrutinizing their data more closely, perhaps we can raise the bar for what's considered an acceptable standard of practice. Moreover, we must strive to shift the emphasis from career-development and promotions to encouraging positive core values, and from large numbers of poor-quality studies to smaller numbers of high-quality ones. A published paper shouldn't be a *REF-returnable output*; it should be an opportunity to contribute to knowledge. A successful grant shouldn't be a resumé-builder; it should be an invitation to conduct some impactful research that will benefit a community. It'll never be a perfect system, but a worn and battered chair that can support our weight is preferable to the one that strains and collapses under the mass of our collective ego.

Holding media accountable for poor science reporting. In the modern context, it isn't enough to do good science and produce a good study; it must have broad impact and garner as much coverage as possible. University publicity and media teams frequently publish press-releases or co-opt the services of a science writer to draw attention to the latest findings of an academic in their department. If the academic is lucky, their institution will employ a dedicated social media representative whose job is to post online links to the latest publication or interview. Unfortunately, it's rare that the media robustly interprets and reports on what can be complex and very esoteric subjects. Consider that much of the media is extrinsically motivated, i.e., interested in a good story. The truth is perceived as boring, unspectacular, and doesn't sell newspapers, magazines, or increase traffic to a website. The articles you read in the press are usually the outliers, the isolated freak results that are rarely replicated by good science. It's the science you scarcely read in the press, reflecting the day-to-day grind of slow incremental progress, that's usually the purveyor of a robust and validated finding. Furthermore, you're much more likely to read of a new product in a glossy online feature than an academic journal, because the former is accessed and shared to a substantially greater magnitude; press-releases with clickbait titles go viral, scientific papers don't. The importance of an objective, unbiased peer-review process is crucial in understanding the scientific principles on which we base our health and performance recommendations, and more emphasis needs to be placed on authentically translating the original findings of this research, and being less tolerant of media articles which distort and misrepresent the message.

One means by which researchers can tackle the misappropriation of study outcomes is by insisting on proof-reading articles or press-releases that contain references to their work, *before* it goes to press. I've always proof-read any article that uses a quote of mine or that cites my research, and I've made a habit of such insistence from the very beginning, even in articles from freelance writers with whom I've worked for years and whom I trust. It isn't that journalists will deliberately misrepresent the research, but they're often not scientists, and just editing a quote for readability – substituting a word, adding a phrase, or omitting punctuation – can change the nuanced meaning of an entire sentence. What a journalist may think is a minor amendment to appease a word count, can change the overall connotation, particularly if it pertains to a complex process or mechanism. The more esoteric a subject, the more likely this is to happen. Researchers can also make a more concerted effort to report their own findings in the media, by composing press-releases or publishing mainstream articles. Many career academics think themselves above writing for non-technical outlets. But if we're not discussing our specialties, and making new knowledge accessible to the general public, then the task is left to writers who don't have the refined appreciation of the science. This is when the waters get muddied and messages lost or misconstrued. Although this won't stem the thirst for likes and shares, scientists are less partial to misrepresenting the facts as they pertain to their own area of study, especially if held to account by their peers.

Better regulation from government/federal bodies on claims made regarding products. Good behavior is learned. Like a petulant child in need of a disciplinarian, product manufacturers and retailers should be held accountable for the claims they make, or the misconduct will simply continue. Perhaps with a greater public emphasis on evidence-based practice, companies will be pressured into providing evidence-for-efficacy alongside their catchy slogans and pretty logos. That's not to say the science will necessarily be rigorous or externally valid, but it'd be a dramatic improvement on the current state of affairs.

In closing, when I reflect on the past several years – since I conceived and began writing this book in 2011 – I note how far I've come as a critical-thinker but, more importantly, how much I still have to learn. With the notion of Socratic Ignorance at the forefront of my mind, I appreciate that learning is an embryonic process, forever reacting to new insights. In the preceding pages, I've done my utmost to frame commercial sport and exercise science objectively, and to help you view it through the critical lens of scientific skepticism. My hope was to challenge the way we conceptualize notions of the health and fitness industry. Throughout the process, several of my own perceptions were tested, and I too was forced to re-evaluate and overturn some preconceived notions of my own and, for that alone, I'm very grateful. My overarching aim was to give something back to a community from which I've already taken so much; I hope, in some small way, I have.

In terms of your path from here, it all depends on the direction in which you'd like to tread. Perhaps subject-specific knowledge is presently paramount. For example, if you're a budding practitioner (e.g., a physiologist, strength and conditioning coach, or psychologist), there's no replacing hard, definitive knowledge of topic. Strive to become an expert in your field, and apply the concepts discussed in this book to complement and stabilize your learning. If you're an athlete or exerciser, perhaps read-up on the supplements you take, the training program you've been following for the past six months but have never reviewed, or the questionable recovery practices you've been following for even longer; there must be sufficient reason (beyond basic routine) for continuation of such practices, so establish their merit for yourself. Expand your knowledge of the practices that form an important component of your day-to-day existence, and see where the rabbit-hole goes. But if this book was your introduction to the various components of critical-thinking, then be aware that there's a whole world of science and skepticism that'll welcome you with open-arms and that's waiting to be explored. You can continue your journey by watching documentaries, YouTube videos, online lectures, TED Talks, or listening to audiobooks or podcasts on the topic; just search the keywords *science, skepticism,* and *critical-thinking.* You could even buy or borrow some of the books I've cited in the preceding chapters. There's no obvious starting point, but if you're overwhelmed by choice, then try *A Demon Haunted World* by Carl Sagan, *Flim Flam* by James Rhandi, *The Skeptic's Guide to the Universe* by Steven Novella, and *The Skeptic's Dictionary* by Robert Todd Carroll. They will all provide you with their own unique perspectives and experiences of this complex and fascinating movement.

In closing, there exists an ongoing battle between critical-thinkers and those that depend on open credulity to line their pockets; fortunately, however, if you're thinking, you're winning. If we are to safely navigate this vast mine-field we call the health and fitness industry, a rudimentary understanding of basic scientific principles will be important, but sharply refined critical-thinking skills will be essential. I hope this book has helped arm you in your ongoing quest for health and fitness or academic progression, and has provided some lessons you can take forward on your journey. If nothing else, I hope to have initiated, provoked, or rekindled a sincere passion to learn about, and understand as best you can, the fascinating natural world in which we make our way. Some of the physical phenomena that influence our lives are bewildering puzzles, but that doesn't mean we should stop trying to elucidate some answers, and it certainly doesn't justify inventing the solutions. The Hitch, as always, gets the last word:

> I want to live my life taking the risk – all the time – that I don't know anything like-enough yet; that I haven't understood enough, that I can't know enough, that I'm always hungrily operating on the margins of a potentially great harvest of future knowledge and wisdom. Take the risk of thinking for yourself; much more happiness, truth, beauty, and wisdom will come to you.
>
> – *Christopher Hitchens*

REFERENCES

Chapter 1: Snake Oil for the 21st-Century

1. Sayers, K., & Menzel, C. (2012). Memory and foraging theory: Chimpanzee utilization of optimality heuristics in the rank-order recovery of hidden foods. *Animal Behaviour, 84*(4), 795–803. doi:10.1016/j.anbehav.2012.06.034
2. Gigerenzer, G., & Gaissmaier, W. (2011). Heuristic decision making. *Annual Review of Psychology, 62*(1), 451–482. doi:10.1146/annurev-psych-120709-145346
3. Marewski, J. N., & Gigerenzer, G. (2012). Heuristic decision making in medicine. *Dialogues of Clinical Neuroscience, 14*(1), 77–89.
4. Kerksick, C. M., Wilborn, C. D., Roberts, M. D., Smith-Ryan, A., Kleiner, S. M., Jager, R., … Kreider, R.B. (2018). ISSN exercise & sports nutrition review update: Research & recommendations. *Journal of The International Society of Sports Nutrition, 15*(38). doi.org/10.1186/s12970-018-0242-y.
5. McClure, S. M., Laibson, D. I., Loewenstein, G., & Cohen, J. D. (2004). Separate neural systems value immediate and delayed monetary rewards. *Science, 306*(5695), 503–507.
6. De Dreu, C. K., Greer, L. L., Handgraaf, M. J., Shalvi, S., Van Kleef, G. A., Baas, M., … Feith, S. W. W. (2010). The neuropeptide oxytocin regulates parochial altruism in intergroup conflict among humans. *Science, 328*(5984), 1408–1411.
7. Heneghan, C., Howick, J., O'Neill, B., Gill, P. J., Lasserson, D. S., Cohen, D., … Thompson, M. (2012). The evidence underpinning sports performance products: A systematic assessment. *British Medical Journal Open, 2*(4). doi:10.1136/bmjopen-2012.

Chapter 3: Logical Fallacies in Sports Science

8. Brooks, G. A. (1986). The lactate shuttle during exercise and recovery. *Medicine and Science in Sport and Exercise, 18*(3), 360–368.
9. Robergs, R. A., Ghiasvand, F., & Parker, D. (2004). Biochemistry of exercise-induced metabolic acidosis. *American Journal of Physiology – Regulatory, Integrative and Comparative Physiology, 287*(3), R502-16.

10. Sinha, R., Cross, A. J., Graubard, B. I., Leitzmann, M. F., & Schatzkin, A. (2009). Meat intake and mortality: A prospective study of over half a million people. *Archives of Internal Medicine, 169*(6), 562–571.
11. Baar, K. (2014). Nutrition and the adaptation to endurance training. *Sports Medicine, 44*(1), S5–12.

Chapter 4: Show Me the Research

12. Cortegiani, A., & Shafer, S. L. (2018). "Think. check. submit." to avoid predatory publishing. *Critical Care (London, UK), 22*(1). doi:10.1186/s13054-018-2244-1.
13. Xia, J., Harmon, J., Connolly, K., Donnelly, R., Anderson, M., & Howard, H. (2015). Who publishes in "predatory" journals? *Journal of the Association For Information Science and Technology, 66*(7), 1406–1417.
14. Song, F., Parekh, S., Hooper, L., Loke, Y. K., Ryder, J., Sutton, A. J., … Harvey, I. (2010). Dissemination and publication of research findings: An updated review of related biases. *Health Technology Assessment (Winchester, UK), 14*(8), iii, ix–xi, 1–193.
15. Keshav, S. (2007). How to read a paper. *ACM SIGCOMM Computer Communication Review, 37*(3), 83–84.
16. Tiller, N. B. (2019). Pulmonary and respiratory muscle function in response to marathon and ultra-marathon running: A review. *Sports Medicine (Auckland, N.Z.), 49*(7), 1031–1041.

Chapter 5: Placebo Products and the Power of Perception

17. Waber, R. L., Shiv, B., Carmon, Z., & Ariely, D. (2008). Commercial features of placebo and therapeutic efficacy. *Jama, 299*(9), 1016–1017.
18. Andersen, J. J. (2018). Expensive running shoes are not better than more affordable running shoes. Retrieved from https://runrepeat.com/expensive-running-shoes-are-not-better-than-more-affordable-running-shoes-study
19. Cannon, W. B. (2002). "Voodoo" death. American anthropologist, 1942;44(new series):169–181. *American Journal of Public Health, 92*(10), 1593–1596; discussion 1594–1595.
20. Montgomery, G., & Kirsch, I. (1996). Mechanisms of placebo pain reduction: An empirical investigation. *Psychological Science,* 7(3), 174.
21. Kong, J., Kaptchuk, T. J., Polich, G., Kirsch, I., & Gollub, R. L. (2007). Placebo analgesia: Findings from brain imaging studies and emerging hypotheses. *Reviews in the Neurosciences, 18*(3–4), 173–190.
22. Petrovic, P., Kalso, E., Petersson, K. M., & Ingvar, M. (2002). Placebo and opioid analgesia-- imaging a shared neuronal network. *Science (New York, N.Y.), 295*(5560), 1737–1740.
23. Bellinger, P. M., Howe, S. T., Shing, C. M., & Fell, J. W. (2012). Effect of combined beta-alanine and sodium bicarbonate supplementation on cycling performance. *Medicine and Science in Sports and Exercise, 44*(8), 1545–1551.
24. Kilding, A. E., Overton, C., & Gleave, J. (2012). Effects of caffeine, sodium bicarbonate, and their combined ingestion on high-intensity cycling performance. *International Journal of Sport Nutrition and Exercise Metabolism, 22*(3), 175–183.
25. Price, M. J., & Cripps, D. (2012). The effects of combined glucose-electrolyte and sodium bicarbonate ingestion on prolonged intermittent exercise performance. *Journal of Sports Sciences, 30*(10), 975–983.
26. Beedie, C. J. (2007). Placebo effects in competitive sport: Qualitative data. *Journal of Sports Science & Medicine, 6*(1), 21–28.

27. Berdi, M., Koteles, F., Hevesi, K., Bardos, G., & Szabo, A. (2015). Elite athletes' attitudes towards the use of placebo-induced performance enhancement in sports. *European Journal of Sport Science, 15*(4), 315–321.
28. Szabo, A., & Muller, A. (2016). Coaches' attitudes towards placebo interventions in sport. *European Journal of Sport Science, 16*(3), 293–300.
29. Clark, V. R., Hopkins, W. G., Hawley, J. A., & Burke, L. M. (2000). Placebo effect of carbohydrate feedings during a 40-km cycling time trial. *Medicine and Science in Sports and Exercise, 32*(9), 1642–1647.
30. Beedie, C. J., Stuart, E. M., Coleman, D. A., & Foad, A. J. (2006). Placebo effects of caffeine on cycling performance. *Medicine and Science in Sports and Exercise, 38*(12), 2159–2164.
31. Garvican, L. A., Pottgiesser, T., Martin, D. T., Schumacher, Y. O., Barras, M., & Gore, C. J. (2011). The contribution of haemoglobin mass to increases in cycling performance induced by simulated LHTL. *European Journal of Applied Physiology, 111*(6), 1089–1101.
32. de Craen, A. J., Kaptchuk, T. J., Tijssen, J. G., & Kleijnen, J. (1999). Placebos and placebo effects in medicine: Historical overview. *Journal of the Royal Society of Medicine, 92*(10), 511–515.
33. Koteles, F., & Ferentzi, E. (2012). Ethical aspects of clinical placebo use: What do laypeople think? *Evaluation & the Health Professions, 35*(4), 462–476.
34. Halson, S. L., & Martin, D. T. (2013). Lying to win-placebos and sport science. *International Journal of Sports Physiology and Performance, 8*(6), 597–599.

Chapter 6: Sports Nutrition

35. Food and Agriculture Organization of the United Nations. (2016). Dietary guidelines for Americans: 2015–2020. Retrieved from https://health.gov/dietaryguidelines/2015/guidelines/
36. Elfhag, K., & Rossner, S. (2005). Who succeeds in maintaining weight loss? A conceptual review of factors associated with weight loss maintenance and weight regain. *Obesity Reviews : An Official Journal of the International Association for the Study of Obesity, 6*(1), 67–85.
37. American Dietetic Association, Dietitians of Canada, American College of Sports Medicine, Rodriguez, N. R., Di Marco, N. M., & Langley, S. (2009). American college of sports medicine position stand. nutrition and athletic performance. *Medicine and Science in Sports and Exercise, 41*(3), 709–731.
38. Bohannon, J. (2015). I fooled millions into thinking chocolate helps weight-loss: Here's how. Retrieved from https://io9.gizmodo.com/i-fooled-millions-into-thinking-chocolate-helps-weight-1707251800
39. Stockwell, T., Zhao, J., Panwar, S., Roemer, A., Naimi, T., & Chikritzhs, T. (2016). Do "moderate" drinkers have reduced mortality risk? A systematic review and meta-analysis of alcohol consumption and all-cause mortality. *Journal of Studies on Alcohol and Drugs,* 77(2), 185–198.
40. Organic Trade Association. (2015). Retrieved from https://www.ota.com/resources/market-analysis
41. Wang, Q., & Sunb, J. (2003). Consumer preference and demand for organic food: evidence from a Vermont survey. Paper presented at the *American Agricultural Economics Association Annual Meeting,* Montreal, Canada.
42. Dangour, A. D., Lock, K., Hayter, A., Aikenhead, A., Allen, E., & Uauy, R. (2010). Nutrition-related health effects of organic foods: A systematic review. *The American Journal of Clinical Nutrition, 92*(1), 203–210.

43. Dangour, A. D., Dodhia, S. K., Hayter, A., Allen, E., Lock, K., & Uauy, R. (2009). Nutritional quality of organic foods: A systematic review. *The American Journal of Clinical Nutrition, 90*(3), 680–685.
44. Smith-Spangler, C., Brandeau, M., Hunter, G., Bavinger, J., Pearson, M., Eschbach, P., … Bravata, D. (2012). Are organic foods safer or healthier than conventional alternatives? A systematic review. *Annals of Internal Medicine, 157*(5), 348–366.
45. Ames, B. N., Profet, M., & Gold, L. S. (1990). Dietary pesticides (99.99% all natural). *Proceedings of the National Academy of Sciences of the United States of America, 87*(19), 7777–7781.
46. Moore, M. C., Cherrington, A. D., Mann, S. L., & Davis, S. N. (2000). Acute fructose administration decreases the glycemic response to an oral glucose tolerance test in normal adults. *The Journal of Clinical Endocrinology and Metabolism, 85*(12), 4515–4519.
47. Uusitupa, M. I. (1994). Fructose in the diabetic diet. *The American Journal of Clinical Nutrition, 59*(3 Suppl), 753S–757S.
48. Cozma, A. I., Sievenpiper, J. L., de Souza, R. J., Chiavaroli, L., Ha, V., Wang, D. D., … Jenkins, D. J. (2012). Effect of fructose on glycemic control in diabetes: A systematic review and meta-analysis of controlled feeding trials. *Diabetes Care, 35*(7), 1611–1620.
49. NHS. (2017). Statistics on obesity, physical activity, and diet. Retrieved from https://digital.nhs.uk/data-and-information/publications/statistical/
50. Lee-Kwan, S. H., Moore, L. V., Blanck, H. M., Harris, D. M., & Galuska, D. (2017). Disparities in state-specific adult fruit and vegetable consumption – United States, 2015. *MMWR.Morbidity and Mortality Weekly Report, 66*(45), 1241–1247.
51. Oyebode, O., Gordon-Dseagu, V., Walker, A., & Mindell, J. S. (2014). Fruit and vegetable consumption and all-cause, cancer and CVD mortality: Analysis of health survey for England data. *Journal of Epidemiology and Community Health, 68*(9), 856–862.
52. World health organization: Obesity and overweight. Retrieved from http://www.who.int/mediacentre/factsheets/fs311/en/
53. Fullfact. Retrieved from https://fullfact.org/news/how-much-does-obesity-cost-nhs/
54. Bates, D., & Price, J. (2015). Impact of fruit smoothies on adolescent fruit consumption at school. *Health Education & Behavior: The Official Publication of the Society for Public Health Education, 42*(4), 487–492.
55. Sense about science: The detox dossier. Retrieved from http://www.senseaboutscience.org/wp-content/uploads/2017/01/Detox-Dossier.pdf
56. Klein, A. V., & Kiat, H. (2015). Detox diets for toxin elimination and weight management: A critical review of the evidence. *Journal of Human Nutrition and Dietetics : The Official Journal of the British Dietetic Association, 28*(6), 675–686.
57. Kim, S., Cha, J. M., Lee, C. H., Shin, H. P., Park, J. J., Joo, K. R., … Choi, J. H. (2012). Rectal perforation due to benign stricture caused by rectal burns associated with hot coffee enemas. *Endoscopy, 44* (Suppl 2) UCTN, E32-E33.
58. Butler, C. C., Vidal-Alaball, J., Cannings-John, R., McCaddon, A., Hood, K., Papaioannou, A., … Goringe, A. (2006). Oral vitamin B12 versus intramuscular vitamin B12 for vitamin B12 deficiency: A systematic review of randomized controlled trials. *Family Practice, 23*(3), 279–285.

Chapter 7: Supplements and Drugs

59. Geyer, H., Parr, M. K., Mareck, U., Reinhart, U., Schrader, Y., & Schanzer, W. (2004). Analysis of non-hormonal nutritional supplements for anabolic-androgenic steroids - results of an international study. *International Journal of Sports Medicine, 25*(2), 124–129.

60. Geyer, H., Parr, M. K., Koehler, K., Mareck, U., Schanzer, W., & Thevis, M. (2008). Nutritional supplements cross-contaminated and faked with doping substances. *Journal of Mass Spectrometry: JMS, 43*(7), 892–902.
61. Gleeson, M., Blannin, A. K., Walsh, N. P., Bishop, N. C., & Clark, A. M. (1998). Effect of low- and high-carbohydrate diets on the plasma glutamine and circulating leukocyte responses to exercise. *International Journal of Sport Nutrition, 8*(1), 49–59.
62. Baar, K., & McGee, S. (2008). Optimizing training adaptations by manipulating glycogen. *European Journal of Sport Science, 8*(2), 97.
63. Fritz, I. B., & McEwan, B. (1959). Effects of carnitine on fatty-acid oxidation by muscle. *Science (New York, N.Y.), 129*(3345), 334–335.
64. Stephens, F. B., Constantin-Teodosiu, D., Laithwaite, D., Simpson, E. J., & Greenhaff, P. L. (2006). Insulin stimulates L-carnitine accumulation in human skeletal muscle. *FASEB Journal: Official Publication of the Federation of American Societies for Experimental Biology, 20*(2), 377–379.
65. Stephens, F. B., Evans, C. E., Constantin-Teodosiu, D., & Greenhaff, P. L. (2007). Carbohydrate ingestion augments L-carnitine retention in humans. *Journal of Applied Physiology (Bethesda, M.d.: 1985), 102*(3), 1065–1070.
66. Witard, O. C., Jackman, S. R., Breen, L., Smith, K., Selby, A., & Tipton, K. D. (2014). Myofibrillar muscle protein synthesis rates subsequent to a meal in response to increasing doses of whey protein at rest and after resistance exercise. *The American Journal of Clinical Nutrition, 99*(1), 86–95.
67. Yang, Y., Breen, L., Burd, N. A., Hector, A. J., Churchward-Venne, T. A., Josse, A. R., … Phillips, S. M. (2012). Resistance exercise enhances myofibrillar protein synthesis with graded intakes of whey protein in older men. *The British Journal of Nutrition, 108*(10), 1780–1788.
68. Areta, J. L., Burke, L. M., Ross, M. L., Camera, D. M., West, D. W., Broad, E. M., … Coffey, V. G. (2013). Timing and distribution of protein ingestion during prolonged recovery from resistance exercise alters myofibrillar protein synthesis. *The Journal of Physiology, 591*(9), 2319–2331.
69. Snijders, T., Trommelen, J., Kouw, I. W. K., Holwerda, A. M., Verdijk, L. B., & van Loon, L. J. C. (2019). The impact of pre-sleep protein ingestion on the skeletal muscle adaptive response to exercise in humans: An update. *Frontiers in Nutrition, 6*, 17.
70. Heneghan, C., Perera, R., Nunan, D., Mahtani, K., & Gill, P. (2012). Forty years of sports performance research and little insight gained. *BMJ (Clinical Research Ed.), 345*, e4797.
71. Bilzon, J. L., Allsopp, A. J., & Williams, C. (2000). Short-term recovery from prolonged constant pace running in a warm environment: The effectiveness of a carbohydrate-electrolyte solution. *European Journal of Applied Physiology, 82*(4), 305–312.
72. EFSA (European Food Safety Authority). (2015). Scientific and technical assistance on food intended for sportspeople. *EFSA Supporting Publication, 12*(9), 32.

Chapter 8: Training Programs and Products

73. Lieberman, D. E. (2012). What we can learn about running from barefoot running: An evolutionary medical perspective. *Exercise and Sport Sciences Reviews, 40*(2), 63–72.
74. Jenkins, D. W., & Cauthon, D. J. (2011). Barefoot running claims and controversies: A review of the literature. *Journal of the American Podiatric Medical Association, 101*(3), 231–246.
75. Jungers, W. L. (2010). Biomechanics: Barefoot running strikes back. *Nature, 463*(7280), 433–434.

76. Divert, C., Mornieux, G., Freychat, P., Baly, L., Mayer, F., & Belli, A. (2008). Barefoot-shod running differences: Shoe or mass effect? *International Journal of Sports Medicine, 29*(6), 512–518.
77. Hanson, N. J., Berg, K., Deka, P., Meendering, J. R., & Ryan, C. (2011). Oxygen cost of running barefoot vs. running shod. *International Journal of Sports Medicine, 32*(6), 401–406.
78. Tam, N., Darragh, I. A. J., Divekar, N. V., & Lamberts, R. P. (2017). Habitual minimalist shod running biomechanics and the acute response to running barefoot. *International Journal of Sports Medicine, 38*(10), 770–775.
79. Hollander, K., Heidt, C., Van Der Zwaard, B. C., Braumann, K. M., & Zech, A. (2017). Long-term effects of habitual barefoot running and walking: A systematic review. *Medicine and Science in Sports and Exercise, 49*(4), 752–762.
80. Warne, J. P., & Gruber, A. H. (2017). Transitioning to minimal footwear: A systematic review of methods and future clinical recommendations. *Sports Medicine - Open, 3*(1), 33.
81. Sigel, B., Edelstein, A. L., Savitch, L., Hasty, J. H., & Felix, W. R., Jr. (1975). Type of compression for reducing venous stasis. A study of lower extremities during inactive recumbency. *Archives of Surgery (Chicago, Ill.: 1960), 110*(2), 171–175.
82. MacRae, B. A., Cotter, J. D., & Laing, R. M. (2011). Compression garments and exercise: Garment considerations, physiology and performance. *Sports Medicine (Auckland, N.Z.), 41*(10), 815–843.
83. Kraemer, W. J., Flanagan, S. D., Comstock, B. A., Fragala, M. S., Earp, J. E., Dunn-Lewis, C., … Maresh, C. M. (2010). Effects of a whole body compression garment on markers of recovery after a heavy resistance workout in men and women. *Journal of Strength and Conditioning Research, 24*(3), 804–814.
84. Beliard, S., Chauveau, M., Moscatiello, T., Cros, F., Ecarnot, F., & Becker, F. (2015). Compression garments and exercise: No influence of pressure applied. *Journal of Sports Science & Medicine, 14*(1), 75–83.
85. Engel, F. A., Holmberg, H. C., & Sperlich, B. (2016). Is there evidence that runners can benefit from wearing compression clothing? *Sports Medicine (Auckland, N.Z.), 46*(12), 1939–1952.
86. Born, D. P., Sperlich, B., & Holmberg, H. C. (2013). Bringing light into the dark: Effects of compression clothing on performance and recovery. *International Journal of Sports Physiology and Performance, 8*(1), 4–18.
87. Barnett, A. (2006). Using recovery modalities between training sessions in elite athletes: Does it help? *Sports Medicine (Auckland, N.Z.), 36*(9), 781–796.
88. Flaherty, G., O'Connor, R., & Johnston, N. (2016). Altitude training for elite endurance athletes: A review for the travel medicine practitioner. *Travel Medicine and Infectious Disease, 14*(3), 200–211.
89. Bonetti, D. L., & Hopkins, W. G. (2009). Sea-level exercise performance following adaptation to hypoxia: A meta-analysis. *Sports Medicine (Auckland, N.Z.), 39*(2), 107–127.
90. Ploszczyca, K., Langfort, J., & Czuba, M. (2018). The effects of altitude training on erythropoietic response and hematological variables in adult athletes: A narrative review. *Frontiers in Physiology, 9*, 375.
91. Chapman, R. F., Stray-Gundersen, J., & Levine, B. D. (1998). Individual variation in response to altitude training. *Journal of Applied Physiology (Bethesda, Md.: 1985), 85*(4), 1448–1456.
92. Porcari, J. P., McLean, K. P., Foster, C., Kernozek, T., Crenshaw, B., & Swenson, C. (2002). Effects of electrical muscle stimulation on body composition, muscle strength, and physical appearance. *Journal of Strength and Conditioning Research, 16*(2), 165–172.
93. Porcari, J. P., Miller, J., Cornwell, K., Foster, C., Gibson, M., McLean, K., & Kernozek, T. (2005). The effects of neuromuscular electrical stimulation training

on abdominal strength, endurance, and selected anthropometric measures. *Journal of Sports Science & Medicine, 4*(1), 66–75.

94. Institute of medicine urges reforms at FDA. (2006). *Lancet (London, UK), 368*(9543), 1211.
95. Richmond, S. J., Brown, S. R., Campion, P. D., Porter, A. J., Moffett, J. A., Jackson, D. A., … Taylor, A. J. (2009). Therapeutic effects of magnetic and copper bracelets in osteoarthritis: A randomised placebo-controlled crossover trial. *Complementary Therapies in Medicine, 17*(5–6), 249–256.
96. Harlow, T., Greaves, C., White, A., Brown, L., Hart, A., & Ernst, E. (2004). Randomised controlled trial of magnetic bracelets for relieving pain in osteoarthritis of the hip and knee. *BMJ (Clinical Research Ed.), 329*(7480), 1450–1454.
97. Dinardi, R. R., de Andrade, C. R., & Ibiapina Cda, C. (2014). External nasal dilators: Definition, background, and current uses. *International Journal of General Medicine, 7*, 491–504.
98. Ottaviano, G., Ermolao, A., Nardello, E., Muci, F., Favero, V., Zaccaria, M., & Favero, L. (2017). Breathing parameters associated to two different external nasal dilator strips in endurance athletes. *Auris, Nasus, Larynx, 44*(6), 713–718.
99. Illi, S. K., Held, U., Frank, I., & Spengler, C. M. (2012). Effect of respiratory muscle training on exercise performance in healthy individuals: A systematic review and meta-analysis. *Sports Medicine (Auckland, N.Z.), 42*(8), 707–724.
100. Karsten, M., Ribeiro, G. S., Esquivel, M. S., & Matte, D. L. (2018). The effects of inspiratory muscle training with linear workload devices on the sports performance and cardiopulmonary function of athletes: A systematic review and meta-analysis. *Physical Therapy in Sport: Official Journal of the Association of Chartered Physiotherapists in Sports Medicine, 34*, 92–104.
101. Alvarez-Herms, J., Julia-Sanchez, S., Corbi, F., Odriozola-Martinez, A., & Burtscher, M. (2019). Putative role of respiratory muscle training to improve endurance performance in hypoxia: A review. *Frontiers in Physiology, 9*, 1970.
102. Shei, R. J., Paris, H. L., Wilhite, D. P., Chapman, R. F., & Mickleborough, T. D. (2016). The role of inspiratory muscle training in the management of asthma and exercise-induced bronchoconstriction. *The Physician and Sportsmedicine, 44*(4), 327–334.
103. Cahalin, L. P., & Arena, R. A. (2015). Breathing exercises and inspiratory muscle training in heart failure. *Heart Failure Clinics, 11*(1), 149–172.
104. Neves, L. F., Reis, M. H., Plentz, R. D., Matte, D. L., Coronel, C. C., & Sbruzzi, G. (2014). Expiratory and expiratory plus inspiratory muscle training improves respiratory muscle strength in subjects with COPD: Systematic review. *Respiratory Care, 59*(9), 1381–1388.

Chapter 9: Complementary and Alternative Therapies in Sport

105. Complementary and alternative medicine market size, share & trends analysis report by intervention (botanical, acupuncture, mind, body, yoga), by distribution (direct contact, E-training), and segment forecasts, 2019–2026. (2019). Retrieved from https://www.grandviewresearch.com/industry-analysis/complementary-alternative-medicine-market
106. Astin, J. A. (1998). Why patients use alternative medicine: Results of a national study. *Jama, 279*(19), 1548–1553.
107. White, J. (1998). Alternative sports medicine. *The Physician and Sportsmedicine, 26*(6), 92–105.
108. Nichols, A. W., & Harrigan, R. (2006). Complementary and alternative medicine usage by intercollegiate athletes. *Clinical Journal of Sport Medicine : Official Journal of the Canadian Academy of Sport Medicine, 16*(3), 232–237.

109. Caspi, O., Koithan, M., & Criddle, M. W. (2004). Alternative medicine or "alternative" patients: A qualitative study of patient-oriented decision-making processes with respect to complementary and alternative medicine. *Medical Decision Making : An International Journal of the Society for Medical Decision Making, 24*(1), 64–79.
110. Lee, M. S., Kim, J. I., & Ernst, E. (2011). Is cupping an effective treatment? an overview of systematic reviews. *Journal of Acupuncture and Meridian Studies, 4*(1), 1–4.
111. Seifman, M. A., Alexander, K. S., Lo, C. H., & Cleland, H. (2017). Cupping: The risk of burns. *The Medical Journal of Australia, 206*(11), 500.
112. Joyce, J., & Herbison, G. P. (2015). Reiki for depression and anxiety. *The Cochrane Database of Systematic Reviews, (4):CD006833.* doi:10.1002/14651858.CD006833.pub2.
113. Thrane, S., & Cohen, S. M. (2014). Effect of reiki therapy on pain and anxiety in adults: An in-depth literature review of randomized trials with effect size calculations. *Pain Management Nursing : Official Journal of the American Society of Pain Management Nurses, 15*(4), 897–908.
114. Gillespie, E. A., Gillespie, B. W., & Stevens, M. J. (2007). Painful diabetic neuropathy: Impact of an alternative approach. *Diabetes Care, 30*(4), 999–1001.
115. Catlin, A., & Taylor-Ford, R. L. (2011). Investigation of standard care versus sham reiki placebo versus actual reiki therapy to enhance comfort and well-being in a chemotherapy infusion center. *Oncology Nursing Forum, 38*(3), E212–20.
116. Manheimer, E., Cheng, K., Wieland, L. S., Min, L. S., Shen, X., Berman, B. M., & Lao, L. (2012). Acupuncture for treatment of irritable bowel syndrome. *The Cochrane Database of Systematic Reviews, (5):CD005111.* doi:10.1002/14651858.CD005111.pub3.
117. Manheimer, E., Cheng, K., Linde, K., Lao, L., Yoo, J., Wieland, S., … Bouter, L. M. (2010). Acupuncture for peripheral joint osteoarthritis. *The Cochrane Database of Systematic Reviews, (1):CD001977.* doi:10.1002/14651858.CD001977.pub2.
118. Sim, H., Shin, B. C., Lee, M. S., Jung, A., Lee, H., & Ernst, E. (2011). Acupuncture for carpal tunnel syndrome: A systematic review of randomized controlled trials. *The Journal of Pain : Official Journal of the American Pain Society, 12*(3), 307–314.
119. Lam, M., Galvin, R., & Curry, P. (2013). Effectiveness of acupuncture for nonspecific chronic low back pain: A systematic review and meta-analysis. *Spine, 38*(24), 2124–2138.
120. Posadzki, P., Zhang, J., Lee, M. S., & Ernst, E. (2012). Acupuncture for chronic nonbacterial prostatitis/chronic pelvic pain syndrome: A systematic review. *Journal of Andrology, 33*(1), 15–21.
121. Kwon, Y. D., Pittler, M. H., & Ernst, E. (2006). Acupuncture for peripheral joint osteoarthritis: A systematic review and meta-analysis. *Rheumatology (Oxford, UK), 45*(11), 1331–1337.
122. Habacher, G., Pittler, M. H., & Ernst, E. (2006). Effectiveness of acupuncture in veterinary medicine: Systematic review. *Journal of Veterinary Internal Medicine, 20*(3), 480–488.
123. Muders, K., Pilat, C., Deuster, V., Frech, T., Kruger, K., Pons-Kuhnemann, J., & Mooren, F. C. (2017). Effects of traumeel (Tr14) on recovery and inflammatory immune response after repeated bouts of exercise: A double-blind RCT. *European Journal of Applied Physiology, 117*(3), 591–605.
124. Schneider, C. (2011). Traumeel - an emerging option to nonsteroidal anti-inflammatory drugs in the management of acute musculoskeletal injuries. *International Journal of General Medicine, 4*, 225–234.
125. Pilat, C., Frech, T., Wagner, A., Kruger, K., Hillebrecht, A., Pons-Kuhnemann, J., … Mooren, F. C. (2015). Exploring effects of a natural combination

medicine on exercise-induced inflammatory immune response: A double-blind RCT. *Scandinavian Journal of Medicine & Science in Sports, 25*(4), 534–542.
126. Lauche, R., Langhorst, J., Lee, M. S., Dobos, G., & Cramer, H. (2016). A systematic review and meta-analysis on the effects of yoga on weight-related outcomes. *Preventive Medicine, 87*, 213–232.
127. Costello, J. T., Baker, P. R., Minett, G. M., Bieuzen, F., Stewart, I. B., & Bleakley, C. (2015). Whole-body cryotherapy (extreme cold air exposure) for preventing and treating muscle soreness after exercise in adults. *The Cochrane Database of Systematic Reviews, (9):CD010789*. doi:10.1002/14651858.CD010789.pub2.
128. Bleakley, C. M., Bieuzen, F., Davison, G. W., & Costello, J. T. (2014). Whole-body cryotherapy: Empirical evidence and theoretical perspectives. *Open Access Journal of Sports Medicine, 5*, 25–36.
129. Rose, C., Edwards, K. M., Siegler, J., Graham, K., & Caillaud, C. (2017). Whole-body cryotherapy as a recovery technique after exercise: A review of the literature. *International Journal of Sports Medicine, 38*(14), 1049–1060.
130. Stump, J. L., & Redwood, D. (2002). The use and role of sport chiropractors in the national football league: A short report. *Journal of Manipulative and Physiological Therapeutics, 25*(3), E2.
131. Ernst, E., & Gilbey, A. (2010). Chiropractic claims in the english-speaking world. *The New Zealand Medical Journal, 123*(1312), 36–44.
132. Ernst, E. (2007). Adverse effects of spinal manipulation: A systematic review. *Journal of the Royal Society of Medicine, 100*(7), 330–338.
133. Todd, A. J., Carroll, M. T., Robinson, A., & Mitchell, E. K. (2015). Adverse events due to chiropractic and other manual therapies for infants and children: A review of the literature. *Journal of Manipulative and Physiological Therapeutics, 38*(9), 699–712.
134. Baldwin, M. L., Cote, P., Frank, J. W., & Johnson, W. G. (2001). Cost-effectiveness studies of medical and chiropractic care for occupational low back pain. a critical review of the literature. *The Spine Journal : Official Journal of the North American Spine Society, 1*(2), 138–147.
135. Hoskins, W., McHardy, A., Pollard, H., Windsham, R., & Onley, R. (2006). Chiropractic treatment of lower extremity conditions: A literature review. *Journal of Manipulative and Physiological Therapeutics, 29*(8), 658–671.
136. Posadzki, P., & Ernst, E. (2011). Spinal manipulation: An update of a systematic review of systematic reviews. *The New Zealand Medical Journal, 124*(1340), 55–71.

Chapter 10: Check Your Ego

137. Strong, R. M. (1902). The metallic colors of feathers from the neck of the domestic pigeon. *Biological Bulletin, 3*(1/2), 85–87.
138. Lord, C. G., Ross, L., & Lepper, M. R. (1979). Biased assimilation and attitude polarization: The effects of prior theories on subsequently considered evidence. *Journal of Personality and Social Psychology, 37*(11), 2098–2109.
139. Garaycoechea, J. I., Crossan, G. P., Langevin, F., Mulderrig, L., Louzada, S., Yang, F., … Patel, K. J. (2018). Alcohol and endogenous aldehydes damage chromosomes and mutate stem cells. *Nature, 553*(7687), 171–177.
140. Twenge, J. M., Martin, G. M., & Spitzberg, B. H. (2018). Trends in U.S. adolescents' media use, 1976–2016: The rise of digital media, the decline of TV, and the (near) demise of print. *Psychology of Popular Media Culture, 8*(4), 329–345.
141. John, L. K., Loewenstein, G., & Prelec, D. (2012). Measuring the prevalence of questionable research practices with incentives for truth telling. *Psychological Science, 23*(5), 524–532.

INDEX